D0364234

Composite Floor Systems

D. L. Mullett

CEng, MICE, MIMechE
Principal Engineer, The Steel Construction Institute

Blackwell Science

The Steel
Construction
Institute

© 1998 D. L Mullett

Blackwell Science Ltd
Editorial Offices:
Osney Mead, Oxford OX2 0EL
25 John Street, London WC1N 2BL
23 Ainslie Place, Edinburgh EH3 6AJ
350 Main Street, Malden
  MA 02148 5018, USA
54 University Street, Carlton
  Victoria 3053, Australia
10, rue Casimir Delavigne
  75006 Paris, France

Other Editorial Offices:

Blackwell Wissenschafts-Verlag GmbH
Kurfürstendamm 57
10707 Berlin, Germany

Blackwell Science KK
MG Kodenmacho Building
7–10 Kodenmacho Nihombashi
Chuo-ku, Tokyo 104, Japan

The right of the Author to be identified as the
Author of this Work has been asserted in
accordance with the Copyright, Designs and
Patents Act 1988

All rights reserved. No part of this publication
may be reproduced, stored in a retrieval system,
or transmitted, in any form or by any means,
electronic, mechanical, photocopying, recording
or otherwise, except as permitted by the UK
Copyright, Designs and Patents Act 1988,
without prior permission of the publisher.

First published 1998

Set in 10/13 pt Times
by Aarontype Limited, Bristol
Printed and bound in Great Britain by
MPG Books Ltd, Bodmin, Cornwall

The Blackwell Science logo is a trade mark of
Blackwell Science Ltd, registered at the United
Kingdom Trade Marks Registry

DISTRIBUTORS
Marston Book Services Ltd
PO Box 269
Abingdon
Oxon OX14 4YN
(*Orders*: Tel: 01235 465500
           Fax: 01235 465555)

USA
Blackwell Science, Inc.
Commerce Place
350 Main Street
Malden, MA 02148 5018
(*Orders*: Tel: 800 759 6102
                  617 388 8250
           Fax: 617 388 8255)

Canada
Login Brothers Book Company
324 Saulteaux Crescent
Winnipeg, Manitoba R3J 3T2
(*Orders*: Tel: 204 224 4068)

Australia
Blackwell Science Pty Ltd
54 University Street
Carlton, Victoria 3053
(*Orders*: Tel: 03 9347 0300
           Fax: 03 9347 5001)

A catalogue record for this title is available from
the British Library

ISBN 0-632-04143-9

Library of Congress
Cataloging-in-Publication Data
Mullett, D. L. (Derek L.)
    Composite floor systems/D. L. Mullett.
      p. cm.
    Includes bibliographical references and index.
    ISBN 0-632-04143-9
    1. Floors – Design and construction.
  2. Composite construction.
  3. Sheet-Steel. I. Title.
  TH2521.M85   1998   97-31335
  690'.16 – dc21   CIP

Although care has been taken to ensure, to the best of our knowledge, that all data and information
contained herein are accurate at the time of publication, the author or the publishers can assume no
liability for any errors in or misinterpretations of such data and/or information or any loss or damage
arising from or related to their use.

1 1575573

Learning Resources
Centre

Contents

Preface   vii

1   Introduction   **1**

2   Composite floor construction   **3**
    2.1   Composite slabs using shallow profiled steel decking   3
        2.1.1   Construction condition   5
        2.1.2   Composite condition   9
        2.1.3   Fire condition   11
        2.1.4   Diaphragm action   11
    2.2   Composite steel beams combined with profiled
          decking and in situ concrete   11
        2.2.1   Construction condition   12
        2.2.2   Effective breadth of slab   12
        2.2.3   Plastic analysis of composite beams   13
        2.2.4   Interaction of shear and moment   15
        2.2.5   Forms of shear connection   15
        2.2.6   Full and partial shear connection   16
        2.2.7   Influence of deck shape   18
        2.2.8   Transverse reinforcement   20
        2.2.9   Elastic section properties   21
        2.2.10 Serviceability criteria   23
        2.2.11 Modular ratios   23
        2.2.12 Deflections   24
        2.2.13 Dynamic sensitivity   25
        2.2.14 Fire condition   25

3   Slimflor construction   **26**
    3.1   Background information   26
    3.2   Slimdek construction   27
    3.3   The general benefits of using Slimdek construction   28
    3.4   The Slimflor beam (SFB)   28
        3.4.1   Slimflor beam combined with the
                composite slab   29
        3.4.2   Construction details   31

	3.4.3	Typical details for internal beams	31
	3.4.4	Tie members	31
	3.4.5	Holes in beam webs	34
3.5	Typical Slimflor project		34
3.6	Design of Slimflor beams using deep decking		36
	3.6.1	Basis of design	36
	3.6.2	Construction stage	37
	3.6.3	Moment resistance including lateral torsional buckling effects	38
	3.6.4	Non-composite beam (Type A)	41
	3.6.5	Section classification	41
	3.6.6	Biaxial stress effects in the flange plate	42
	3.6.7	Construction stage effects	44
	3.6.8	Vertical shear resistance	47
	3.6.9	Moment resistance of non-composite beam (laterally restrained)	47
	3.6.10	Composite beam (Type B)	48
	3.6.11	Moment capacity of composite beam	49
	3.6.12	Serviceability stresses	52
	3.6.13	Transverse reinforcement	53
	3.6.14	Natural frequency of the beam	54
	3.6.15	Edge beams	54
	3.6.16	Connections	55
3.7	Slimflor beam test programme		55
	3.7.1	Fire resistance	55
	3.7.2	Slimflor beam test	56
3.8	Asymmetric Slimflor Beam (ASB)		60
	3.8.1	Introduction	60
	3.8.2	ASB section sizes	61
	3.8.3	Design principles	64
	3.8.4	Ultimate limit state	65
	3.8.5	Serviceability limit state	67
	3.8.6	Fire limit state	68
	3.8.7	Design procedures for ASB sections	69
	3.8.8	Moment resistance of composite section	69
	3.8.9	Full shear connection	70
	3.8.10	Longitudinal force in the slab	72
	3.8.11	Partial shear connection	73
	3.8.12	Transverse reinforcement	74
	3.8.13	Elastic properties	76
3.9	RHS Slimflor edge beams (RHSFB)		77
	3.9.1	Edge beams in Slimflor construction	77
	3.9.2	Form of construction	78

3.9.3   RHS Slimflor edge beams acting
        non-compositely with floor slab   78
3.9.4   RHS Slimflor edge beams acting compositely
        with floor slab   79
3.9.5   Construction details   80
3.9.6   Support of cladding   85
3.9.7   Design of RHS Slimflor edge beams   89
3.9.8   Torsional effects on the RHS edge beam   90
3.9.9   Design of non-composite beams (Type A)   91
3.9.10 Design of composite beams (Type B)   97
3.9.11 Fire resistance design   102
3.9.12 General requirements of BS 5950: Part 8   102
3.9.13 Additional fire protection   103
3.10 Composite slabs using deep decking (SD225)   103
        3.10.1   Construction stage   104
        3.10.2   Section properties   107
        3.10.3   Composite slab design   108
        3.10.4   Installation notes   110

4   **Other forms of composite construction**   **112**
    4.1   Composite trusses   112
          4.1.1   Introduction   112
          4.1.2   Composite truss systems   113
          4.1.3   Truss member types   118
          4.1.4   Hand analysis of composite trusses   120
          4.1.5   Typical detailing   128
    4.2   Fabricated composite beams   129
          4.2.1   General description of structural arrangements and
                  forms using fabricated sections   129
          4.2.2   Accommodation of services   132
    4.3   Stub girders   134
          4.3.1   Introduction   134
          4.3.2   Design considerations   136
          4.3.3   Design procedure for stub girders   153
          4.3.4   Conclusions   155
    4.4   The parallel beam approach   156
          4.4.1   Conventional approach   156
          4.4.2   Parallel beam approach   157
          4.4.3   Detailed aspects of parallel beam framing   159
          4.4.4   Construction aspects   162
          4.4.5   Fabrication   162
          4.4.6   Erection   162
          4.4.7   Services   163

4.4.8   Site welding   164
4.4.9   Lateral stability of frame and beams   164
4.4.10 Propping during construction   165
4.5   Cellular beams   165
4.5.1   Introduction   165
4.5.2   Benefits of cellular beam construction   168
4.5.3   Beam geometry   168
4.5.4   Examples of use   169

**Appendix A   Composite construction using shallow profiled
                  steel decking**   **171**
A1   Derivation of formulae used for composite
       construction with shallow profiled steel decking   171
A2   Worked example for a typical 10 m span composite
       beam subject to a uniformly distributed load   185

Appendix B   Slimdek construction   **201**
B1   Derivation of formulae   201
B1.1   Derivation of formulae for the Slimflor
         beam (SFB) sections   201
B1.2   Derivation of design formulae for
         Asymmetric Slimflor Beam (ASB) sections   215
B1.3   Formulae for plastic moment resistance of
         RHS Slimflor edge beam   225
B2   Worked examples   233
B2.1   Non-composite Slimflor beam (Type A)   233
B2.2   Composite Slimflor beam (Type B)   257

Appendix C   Worked example for stub girder construction **281**

References   307

Index   309

Preface

The methods of construction described in this book reflect the development and use of steel-composite construction for buildings in the UK over the past 15 to 20 years. The author has been involved with The Steel Construction Institute (SCI) design guides covering the topic of composite floor systems and, before the SCI, the Constructional Steel Research and Development Organisation (Constrado). Extracts from SCI design guides, covering text and worked examples, have been used in the preparation of this book. They are reproduced with the agreement of the Director of the SCI and the authors, Dr R. M. Lawson, Dr J. W. Rackham, Dr G. W. Owens, Dr R. McConnel, Mr J. Rushton and Mr S. Neil.

Two basic concepts of construction have been identified:

- Composite steel beams combined with profiled decking and in situ concrete
- Slimdek construction.

The first concept identified above is well established in the UK as the traditional approach to forming a composite floor. It is now accepted that the introduction of this method of construction dramatically increased the market share for steel framed multi-storey buildings in the UK.

Slimdek construction is essentially where the floor slab is formed within the depth of the supporting beam. This approach was developed initially for the low-rise market, as a result of the demands by specifiers to reduce the construction depth.

These two basic forms of construction are generally considered for spans up to 10 m. In addition, Chapter 4 has been included which deals with longer spanning methods for forming a composite floor.

This book would not have been possible without the funding and assistance of British Steel plc for the research and development of the forms of construction presented here. The author is also grateful to his colleague Dr R. M. Lawson of the SCI for kindly agreeing to edit this publication.

Slimflor® and Slimdek® are registered trademarks of British Steel plc.

Chapter 1
Introduction

The author's involvement with composite construction started some 20 years ago when a joint decision between the Constructional Steel Research and Development Organisation (Constrado) and British Steel was made to improve the market share for steel framed multi-storey buildings. In the UK at that time, steel framed multi-storey buildings had a modest share of this market, mainly because of the outdated methods of steel construction that were available. Reinforced concrete structures were dominant and few engineers felt competent in steel or composite construction design.

A well known property developer based in the UK had visited North America and reported that a new method of composite construction was proving to be highly successful, especially in terms of speed of construction. The key features were the use of steel decking, both as permanent formwork and reinforcement to an in situ concrete slab, and the ability to weld shear connectors through the decking on site.

Having been informed of this new method of construction, engineers from Constrado and British Steel arranged feasibility studies to seek the opinion of architects, consulting engineers, property developers and decking manufacturers to gauge how UK industry would receive such a radical idea. A great deal of enthusiasm was generated by all concerned. To further endorse this form of construction, a team of engineers and architects were sent to North America on a fact-finding exercise.

The team were suitably impressed with seeing the form of construction in use and the 'fastrack' methods adopted for the construction programme. Nevertheless, it soon became evident that the major limitation was the lack of design guidance. In particular, the Code of Practice CP117 (which has been replaced by BS 5950: Part 3) did not cover composite construction using profiled steel decking. To overcome this problem, Constrado was commissioned to produce an authoritative document to include design procedures, typical construction details and a set of comprehensive load tables that could be used for rapid sizing of the steel members.

The task of preparing the design procedures was this author's responsibility. The publication was entitled *Design recommendations for composite floors and beams using steel decks – Section 1 Structural* and was used for approximately 10 years. This publication was also used by a number of computer-based companies to prepare the first composite beam design programs.

It is now common knowledge that this method of construction dramatically increased the market share for steel framed multi-storey buildings. However, in recent years, reinforced concrete 'flat slab' construction has tended to dominate the low-rise market. This is partly due to a recession and the servicing benefits that can be gained using 'flat slab' construction.

To restore the balance in steel's favour, British Steel commissioned the Steel Construction Institute (SCI) to design and develop the Slimflor® concept for use in the UK. This is effectively a flat floor in which the steel section is partially encased in concrete which enhances its fire resistance.

The Slimflor form of construction has been described as the natural successor to the method of composite construction that uses profiled steel decking (40–70 mm deep) spanning between the beams.

In such a relatively short period of time the system has attracted a great deal of attention. The Slimflor concept has been successfully used on a number of projects in the UK and in Europe, and has attracted worldwide attention.

The traditional forms of composite construction that use shallow profiled steel decking and Slimflor construction are generally used in the short-to-medium span range. The other methods of construction described in this book concentrate mainly on long-span solutions (up to 25 m). Worked examples and the derivation of formulae are included covering the most popular forms of construction.

Chapter 2
Composite Floor Construction

2.1 Composite slabs using shallow profiled steel decking

The design of steel decking and composite slabs is covered by BS 5950: Part 4[1] which was the first part of BS 5950 to be used in 1982 and was influential in developing this form of construction.

Composite slabs comprise profiled steel decking as the permanent formwork to the underside of concrete slabs spanning between support beams. The decking acts compositely with the concrete under service loading. It also supports the loads applied to it before the concrete has gained adequate strength. A light mesh reinforcement (generally A142) is placed in the concrete slab.

The decking performs a number of roles and is an integral part of the structural system:

- It supports loads during construction
- It acts as a working platform and protects workers below
- It develops composite action with the concrete to resist the imposed loading on the floor
- It transfers in-plane loads by diaphragm action to vertical bracing or shear walls
- It stabilises the beams against lateral torsional buckling
- It can act as transverse reinforcement to the composite beams.

The main economy sought in buildings is speed of construction, and for this reason slabs and beams are usually designed to be unpropped during the construction stage. Spans of the order of 3.0–4.0 m between support beams are common, and beams are usually designed to span between 6 m and 12 m. Connections between the structural steel elements are generally designed as 'simple', i.e. not moment-resisting. The main elements of construction of typical composite building are illustrated in Figs 2.1 and 2.2.

Figure 2.1 Typical composite steel-framed building during construction.

Figure 2.2 Composite slab during concreting.

2.1.1   *Construction condition*

The design of composite slabs is often limited by the ability of the deck to resist the loads applied to it during construction where the deck is not temporarily propped. The following aspects of designs in the construction stage should be noted.

Deck types

Modern deck profiles are in the range of 45–80 mm in height and 150–300 mm trough spacing. There are two well known generic types: the re-entrant (dovetail) profile and the trapezoidal profile with web indentations. Typical profiles are shown in Figs 2.3 and 2.4.

Steel grades and thicknesses

Galvanised sheet steel for this application is typically 0.9–1.5 mm thick. Z28 steel (280 N/mm^2 yield strength) is generally specified, although Z35 steel (350 N/mm^2 yield strength) is used for some of the deeper, longer-span profiles. The thickness of galvanising is usually approximately 0.02 mm per face, equivalent to 275 g/m^2 total coverage.

Slab span and depths

The most efficient use of composite slabs is for spans between 3.0 m and 4.0 m. Slab depths largely depend on insulation requirements in fire and are usually between 100 mm and 150 mm. For most designs

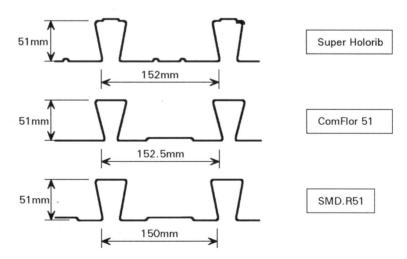

Figure 2.3   Dovetailed deck profiles used in composite slabs.

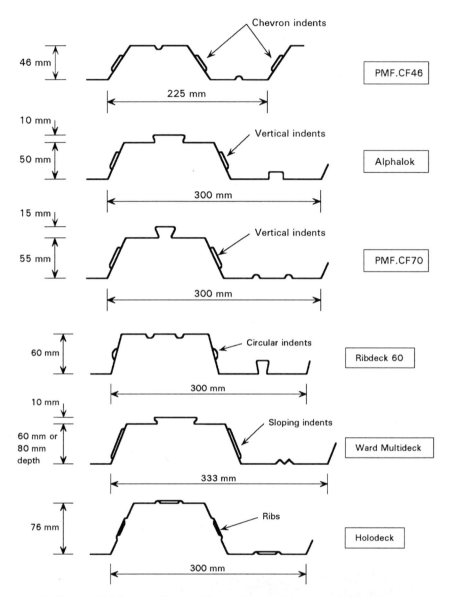

Figure 2.4 Trapezoidal deck profiles used in composite slabs.

the slab span-to-depth ratio should not exceed the limits given in the section below on serviceability.

Concrete type and grade

Normal (NWC) and lightweight (LWC) concretes are both used. The modern method of placement is by pump. The dry density of structural lightweight concrete (Lytag and sand aggregate) is 1750–1850 kg/m³. The wet density is used when determining the loading on the decking in the construction stage and is typically 100 kg/m³ greater than the dry density. The concrete grade (cube strength in N/mm²) is specified as 30–40. The concrete type affects the strength of the shear connectors.

Construction loading

The decking supports the weight of the concrete in the finished slab, the excess concrete arising from the deflection of the decking (ponding), the weight of the operatives and any impact loads. The construction load (in addition to the self-weight of the slab) is currently specified as a uniform load of 1.5 kN/m² when designing the decking. This should not be confused with the construction load of 0.5 kN/m² when designing the composite beam. The form of loading developed during the concreting operation is illustrated in Fig. 2.2.

Strength design of decking

The design of continuous decking (spanning over a number of beams) is based, according to BS 5950: Part 4[1], on an elastic distribution of

Figure 2.5   Local buckling of composite deck in bending.

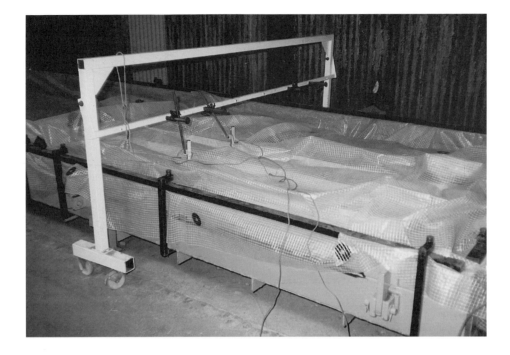

Figure 2.6   Deck testing using the vacuum rig at British Steel's Welsh Technology Centre, Port Talbot.

moment, as a safe lower bound to the collapse strength. Elastic moments are greatest at the supports. The flexural resistance of the section is also determined elastically, as its performance is limited by local buckling (Fig. 2.5). The strength of the section can be improved by introducing stiffening folds in the compression elements.

The elastic approach is relatively conservative. Tests demonstrate a redistribution of moment from the highly stressed areas at the supports to the mid-span area. Spans 10 to 15% in excess of those predicted by elastic design are possible whilst maintaining an adequate factor of safety against failure and satisfactory performance at working load. It is for this reason that manufacturers often present load-span tables on the basis of tests rather than the elastic approach in BS 5950: Part 4[1]. A typical vacuum rig test arrangement is shown in Fig. 2.6.

Deflection limits

A limit on the residual deflection of the soffit of the deck (after concreting) of span/180 is specified, increasing to span/130 if the effects of 'ponding' are included in the strength and deflection assess-ment of the decking. In the second case the effect of the increased weight of concrete should be included in the design of the support structure.

2.1.2   *Composite condition*

Composite slabs are usually designed as simply supported elements with no account taken of the continuity provided by the slab reinforcement at ultimate loads. Composite slabs fail by incomplete shear connection when end anchorage is not provided. This means that failure occurs by slip between the deck and the concrete before the plastic bending capacity of the slab is reached.

Modes of failure

The ultimate bending resistance of composite slabs (in the absence of end anchorage) is controlled by slip between the concrete and the deck. This occurs by a combination of friction and chemical bond between the concrete and the deck, followed by mechanical interlock after initial slippage has taken place. This is known as 'shear bond' failure. In unpropped construction, only the loads applied to the composite section after concreting are considered.

If adequate end anchorage is provided then the composite slab can reach its bending resistance as a reinforced concrete slab in which the area of decking acts as conventional reinforcement to the concrete. The vertical shear resistance is also assessed as a reinforced concrete slab.

Mechanical interlock

Mechanical interlock between the concrete and the deck is improved by the use of indentations or embossments in the webs of the deck, illustrated in Figs 2.3, 2.4 and 2.5. The dovetail profiles in Fig. 2.3 achieve their shear bond capacity largely by preventing separation of the deck and the concrete.

Design by testing

In view of the large number of interacting parameters influencing the strengths, the performance of a particular composite slab can only be readily assessed by tests of the form shown in Fig. 2.7. According to BS 5950: Part 4[1], a minimum of six slabs are to be tested covering a range of design parameters. The slabs are first subject to dynamic

Figure 2.7   Standard test on composite slab.

loading between 50% and 150% of the desired working load, and then the load is increased statically to failure. The objective of the dynamic part of the test is to identify those cases where there is an inherently fragile connection between the concrete and the deck.

The test information is used in general design by the establishment of empirical constants broadly defining the mechanical interlock and friction bond components of shear resistance. Because of the empirical nature of the design formula in BS 5950: Part 4, manufacturers normally present this information in the form of load–span tables. It is generally found that the degree of composite action is sufficiently high for most designs to be controlled by the construction condition.

Serviceability

The key serviceability aspects are the avoidance of premature slip between the deck and the concrete slab and the control of deflections. Although first slip occurs well below the ultimate load, it rarely causes a serviceability problem in well designed composite slabs. Therefore, the deflection of a composite slab can be assessed from its stiffness as a reinforced concrete slab.

As a general rule, span:depth ratios (based on the overall slab depth) of continuous composite slabs should be less than 30 for LWC and 35 for NWC in order to satisfy normal deflection limits. Continuity at serviceability is provided by mesh reinforcement. For single-span slabs these ratios should be reduced to 25 for LWC and 30 for NWC. Deflections should be calculated for designs outside these limits.

2.1.3 *Fire condition*

In principle, the minimum depth of the composite slab is controlled by 'insulation' requirements in a fire, and the amount of reinforcement may be determined from a 'fire engineering' analysis of the reduced strength of the slab subject to elevated temperatures. Slabs in LWC are thinner than those in NWC because of the better insulating properties of the aggregate.

The adequacy of composite slabs for up to 90 minutes fire resistance has been established, and this is now covered by BS 5950: Part 8.[2]

2.1.4 *Diaphragm action*

In-plane forces are developed in the decking during construction and in the slab in service. It is assumed that the decking is stiff and strong enough to act as a shear diaphragm provided the decking is attached by shear connectors to the steel beams. Temporary strength is provided by edge fasteners spaced at not more than 600 mm apart. It is not necessary to use seam fasteners between the sheets except in extreme circumstances where high shear forces are to be resisted.

In the composite stage, the slab is often designed to transfer relatively high in-plane forces to cores or vertical bracing. These forces are developed in the slab via the shear connectors. It is not normally necessary to provide additional shear connectors except close to points of local transfer of shear force, e.g. adjacent to vertical bracing.

2.2 Composite steel beams combined with profiled decking and in situ concrete

Composite construction in buildings in the context of BS 5950: Part 3: Section 3.1[3] means the design of steel beams, usually of I section, to act 'compositely' with concrete or composite slabs by use of shear connectors. Composite action is responsible for a considerable increase in the load-carrying capacity and stiffness of steel beams, which when utilised in design, can result in significant savings in steel weight and in construction depth. These economies have largely accounted for the dominance of composite steel frame construction in the commercial building sector in the UK in recent years.

The structural system of a composite beam is essentially a series of parallel T beams with thin wide flanges. The concrete flange is in compression and the steel beam is largely in tension. The benefits of composite action in terms of strength and serviceability are considerable, leading to economy in the sizing of the steel beams.

The bending resistance of the section is evaluated on 'plastic' analysis principles, whereas the serviceability performance is evaluated on elastic section analysis principles. S355 (grade 50) steel is often preferred for steel beams, which are usually designed to be unpropped during construction. Lightweight concrete often proves to be more economic than normal weight concrete.

Where simply supported unpropped composite beams are sized on the basis of their plastic capacity it is normally found that span-to-depth ratios can be in the range of 18–22 before serviceability criteria influence the design of the beam. The 'depth' in these cases is defined as the overall depth of the floor (beam and slab).

2.2.1 Construction condition

In unpropped construction, the steel beam is sized first to support the self-weight of the slab and other construction loads before the concrete has gained adequate strength for composite action. The construction load (treated as an imposed load) should be taken as not less than $0.5\,\text{kN/m}^2$.

Beams are assumed to be laterally restrained by the decking in cases where the decking spans perpendicular to the beams and is directly attached to them. Such beams can develop their full flexural capacity. In cases where the decking spans parallel to the beams, lateral restraint is offered only at the transverse connections of secondary beams to the primary beams. The bending resistance of the beams can be assessed from BS 5950: Part 1[4] using the appropriate slenderness between restraints.

2.2.2 Effective breadth of slab

The effective breadth of the slab is not a precise figure, as it depends on the form of loading and the position along the beam. For compatibility between designs at the ultimate and serviceability limit states, the effective breadth is taken as span/4 for internal beams (divided equally between each side of the beam) but not exceeding the actual slab width (distance between beams) considered to act with each beam.

2.2.3 *Plastic analysis of composite beams*

The ultimate bending strength of a composite section is determined from its plastic moment capacity. For ease of calculation, rectangular stress blocks are used as opposed to the actual case, see Fig. 2.8.

The plastic moment capacity of the section is independent of the order of loading, i.e. propped or unpropped construction. The plastic moment of resistance is compared to the moment resulting from the total factored loading using the load factors in BS 5950: Part 1.[4]

The plastic neutral axis of the composite section is evaluated assuming stresses of p_y in the steel (determined from BS 5950: Part 1) and $0.45f_{cu}$ in the concrete. The tensile resistance of the steel is, therefore, $R_s = p_yA$, where A is the cross-sectional area of the beam. The compressive resistance of the concrete slab depends on the orientation of the decking. Where the decking crosses the beams, the depth of concrete contributing to the compressive capacity is $D_s - D_p$ (Fig. 2.9(a)). Clearly, D_p is zero in a solid slab. Where the decking runs parallel to the beams, then the total cross-sectional area of the concrete may be used (Fig. 2.9(b)). However, for convenience, the concrete within the deck troughs is usually ignored. Hence, the compressive resistance of the concrete slab is

$$R_c = 0.45f_{cu}(D_s - D_p)B_e$$

where B_e is the effective breadth of the slab.

Three cases of plastic neutral axis (PNA) depth y_p (measured from the upper surface of the slab) exist. These are presented in Fig. 2.10. It

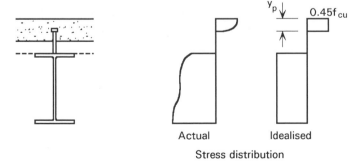

Actual   Idealised

Stress distribution

Figure 2.8   Plastic analysis of composite section.

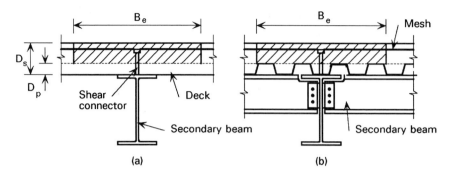

Figure 2.9 Composite beams incorporating composite deck slabs. (a) Deck perpendicular to secondary beam. (b) Deck parallel to primary beam.

is not necessary to calculate y_p explicitly if the following formulae for the plastic moment resistance of I section beams subject to positive (sagging) moment are used. R_w is the axial capacity of the web and R_f is the axial capacity of one steel flange (the section is assumed to be symmetrical). The top flange is considered to be fully restrained by the concrete slab.

Case 1: $R_c > R_s$ (plastic neutral axis lies in concrete slab)

$$M_{pc} = R_s\left[\frac{D}{2} + D_s - \frac{R_s}{R_c}\left(\frac{D_s - D_p}{2}\right)\right]$$

Case 2: $R_s > R_c > R_w$ (plastic neutral axis lies in steel flange)

$$M_{pc} = R_s\frac{D}{2} + R_c\left(\frac{D_s - D_p}{2}\right) - \frac{(R_s - R_c)^2}{R_f}\frac{T}{4}$$

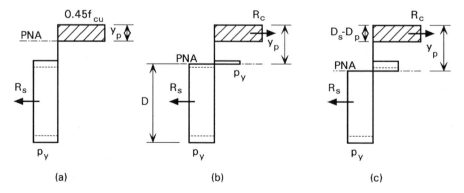

Figure 2.10 Plastic analysis of composite section under positive moment. (a) y_p in slab, (b) y_p in steel flange, (c) y_p in steel web.

Note that the last term in this expression is generally small (T is the flange thickness).

Case 3: $R_c > R_w$ (plastic neutral axis lies in steel web)

$$M_{pc} = M_s + R_c \left(\frac{D_s + D_p + D}{2} \right) - \frac{R_c^2}{R_w} \frac{D}{4}$$

where M_s is the plastic moment resistance of the steel section alone.

This formula assumes that the web is compact (i.e. not subject to the effects of local buckling). In this case, the depth of the web in compression should not exceed $73t\epsilon$, where t is the web thickness (ϵ is defined as $\sqrt{(275/p_y)}$ in BS 5950: Part 1). If the web is non-compact, a method of determining the capacity of the section is given in BS 5950: Part 3; Section 3.1, Appendix B (see Appendix A1 for further information).

2.2.4 *Interaction of shear and moment*

Vertical shear can cause a reduction in the plastic moment resistance of a composite beam. This only occurs where high moment and shear co-exist at the same position along the beam (i.e. the beam is subject to one or two point loads). Where the shear force F_v exceeds $0.5P_v$ (where P_v is the lesser of the shear resistance and the shear buckling capacity, determined from BS 5950: Part 1), the reduced moment capacity should be determined from

$$M_{cv} = M_c - (M_c - M_f) \left(\frac{2F_v}{P_v} - 1 \right)^2$$

where: M_c is the plastic moment resistance of the composite section
M_f is the plastic moment resistance of the composite section having deducted the shear area (the web) of the section.

2.2.5 *Forms of shear connection*

The modern form of welded shear connection is the headed stud. The most popular size is 19 mm diameter and 100 mm height. Studs are often welded through the decking using a hand tool connected via a control unit to a power generator (see Fig. 2.11). There are, however, some limitations to through-deck welding: firstly, the top flange of the beam should not be painted or, alternatively, the paint should be removed from the zone where the shear connectors are to be welded; secondly, the galvanised steel should be less than 1.25 mm thick and should be clean and free from moisture.

Figure 2.11   Through-deck welding of shear stud connector.

All shear connections should be capable of resisting uplift forces caused by the tendency for the slab to separate from the beam. Hence, headed rather than plain studs are used.

The strength of shear connectors is a function of the concrete strength and type, and is determined from the standard push-out test. Characteristic strengths of stud shear connectors are given in BS 5950: Part 3: Section 3.1.[3] The ultimate tensile strength of the steel used in the shear connectors (before forming) should be not less than 450 N/mm^2 and the elongation at failure not less than 15%.

The design strength of stud shear connectors is taken as 80% of their characteristic strength. This is to ensure that flexural failure of the beam occurs in preference to longitudinal shear failure. A further 10% reduction in strength is made where lightweight concrete is used (density > 1750 kg/m^3).

2.2.6 *Full and partial shear connection*

In simple composite beams subject to uniform load, the elastic shear flow defining the shear transfer between the slab and the beam is linear, increasing to a maximum at the ends of the beam. Beyond the elastic limit of the shear connectors there is a transfer of force along the beam such that, at failure, each of the shear connectors is assumed to resist equal force. This implies that the shear connectors possess adequate deformation capacity.

In the plastic design of composite beams, the longitudinal shear force to be transferred between the points of zero and maximum moment should be the lesser of R_c or R_s (see Section 2.2.3). If so, full shear connection is provided.

In cases where fewer shear connectors than the number required for full shear connection are provided, it is not possible to develop the full plastic moment capacity of the section. The stress block method, as in Section 2.2.3, may be modified to take into account the effects of 'partial shear connection'. The design formulae are given in Appendix A1. The degree of shear connection may be defined as

$$K = \frac{R_q}{R_s} \qquad \text{for } R_s < R_c$$

$$\text{or} \quad K = \frac{R_q}{R_c} \qquad \text{for } R_c < R_s$$

where R_q is the total shear force transferred by the shear connectors between the points of zero and maximum moment.

Traditionally, the moment capacity of a composite section can be defined in terms of a linear interaction with the degree of shear connection K, such that

$$M_c = M_s + K(M_{pc} - M_s)$$

where M_s is the plastic moment capacity of the steel section
       M_{pc} is determined as in Section 2.2.3.

The stress-block and linear interaction methods are compared in Fig. 2.12. Clearly, the stress-block method offers some benefit in terms of moment capacity. However, the advantage of the linear interaction method is that the different spacings of shear connectors can be assessed readily.

In using these methods a lower limit of K of 0.4 is specified. This is to avoid any adverse effects arising from the limited deformation capacity of the shear connectors. The slip at the ends of a composite beam increases with span and inversely with the degree of shear

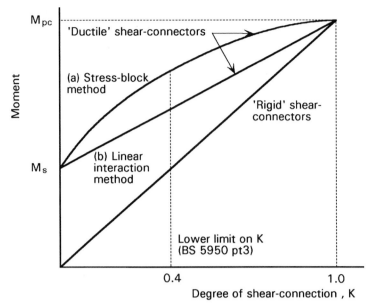

Figure 2.12   Interaction between moment capacity and degree of shear connection in composite beams.

connection. In BS 5950: Part 3: Section 3.1, the minimum degree of shear connection increases with span (L metres) such that

$$K \geq \frac{L-6}{10} \geq 0.4$$

This formula means that beams longer than 16 m are to be designed for full shear connection. In principle, the above equation is intended to avoid excessive deformation of the shear connectors of long-span beams.

Also partial shear connection is not permitted for beams subject to heavy off-centre point loads, where the position of maximum moment is close to the support. A further requirement is that the moment capacity of beams subject to point loads should be adequate at all locations along the beam. It may be necessary to check the shear connection provided at intermediate points or, alternatively, distribute the shear connectors in accordance with the shear force diagram.

2.2.7 Influence of deck shape

The efficiency of the shear connection between the composite slab and the composite beam may be reduced as a result of the shape of the deck. The reduction factor, k, which may be defined as shown below, multiplies the design resistance of the shear connectors.

Decking perpendicular to beam:

(1) For one shear connector per trough

$$k = 0.85\left(\frac{b_r}{D_p}\right)\left(\frac{h}{D_p} - 1\right) \le 1.0$$

(2) For two shear connectors per trough

$$k = 0.6\left(\frac{b_r}{D_p}\right)\left(\frac{h}{D_p} - 1\right) \le 0.8$$

where b_r is the average trough width (for trapezoidal deck profiles) or minimum trough width (for re-entrant deck profiles)
h is the as-welded height of the shear connector ($h \ge D_p + 35$ but $h \le 2 D_p$ in the above equations)
D_p is the deck profile height.

Restrictions are placed on the dimension b_r that may be used in the above equations when shear connectors are welded non-centrally in the troughs. For shear connectors placed in the 'unfavourable' location, b_r should be taken as $2e$, where e is the minimum distance from the centre of the shear connector to the mid-height of the adjacent web of the profile (or to the top of the web in re-entrant profiles). These cases are illustrated in Fig. 2.13. For shear connectors placed in pairs, but in an off-set pattern, alternately on the unfavourable and favourable sides of the trough, b_r may be determined as for centrally located shear connectors.

Decking parallel to beam:

For $b_r/D_p \ge 1.5$, $k = 1.0$

For $b_r/D_p < 1.5$,

$$k = 0.6\left(\frac{b_r}{D_p}\right)\left(\frac{h}{D_p} - 1\right) \le 1.0$$

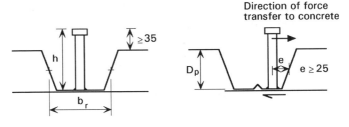

Trapezoidal profiles

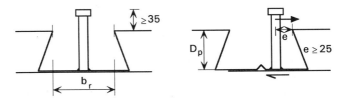

Re-entrant profiles

Figure 2.13   Geometrical parameters used in determining reduction factors due to deck shape.

2.2.8 *Transverse reinforcement*

The longitudinal shear strength of the concrete slab should be checked, in order to ensure transfer of force from the shear connectors into the slab without splitting the concrete. This may require provision of transverse reinforcement (perpendicular to the beam). Potential shear planes through the slab lie on either side of the shear connectors

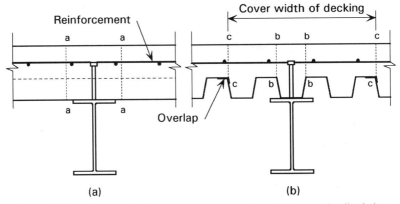

Figure 2.14   Potential failure planes through the slab in longitudinal shear. (a) Deck perpendicular to beam. (b) Deck parallel to beam.

(Fig. 2.14). The shear resistance per unit length of shear plane along the beam is

$$V_c = 0.03\eta f_{cu} A_{cv} + 0.7 A_r f_y \leq 0.8\eta A_{cv} \sqrt{f_{cu}}$$

where A_{cv} is the cross-sectional area of concrete per unit length in any shear plane
A_r is the amount of the reinforcement crossing each shear plane
f_y is the yield strength of the reinforcement
η is taken as 1.0 for NWC and 0.8 for LWC.

To this longitudinal shear resistance may be added a component arising from the tensile strength of the deck. Its full strength can be used when the deck crosses the beams (i.e. secondary beams) and is continuous. There are situations where the deck is discontinuous. In such cases, the anchorage force developed by the shear connectors may be included, provided both ends of the deck are properly attached. (However, it is general practice to ignore the deck contribution in such cases.) The anchorage force is given by

$$V_p = n\,(4\,\phi\,t_s\,p_{ys})$$

where n is the number of shear connectors per unit length connecting each sheet
ϕ is the stud diameter
t_s is the sheet thickness
p_{ys} is the design strength of the sheet steel.

The contribution of the decking should be neglected where it is not properly anchored (i.e. at discontinuities or at edge beams) or where sheet overlaps are present close to the beam. This is often the case in primary beams where the deck is placed parallel to the beams. In such cases additional transverse reinforcement is often required in the slab. It is usually found that mesh reinforcement provides adequate transverse reinforcement in the design of secondary beams.

2.2.9   Elastic section properties

The important elastic properties of the section are the section modulus and the second moment of area. It is first necessary to determine the centroid (elastic neutral axis) of the transformed section by expressing the area of concrete in steel units. This is done by dividing the concrete area within the effective breadth of the slab, B_e, by an appropriate modular ratio, α_e (ratio of the elastic modulus of steel to concrete determined as in Section 2.2.11).

In unpropped construction, account should be taken of the stresses induced in the non-composite section as well as the stresses in the composite section. In elastic analysis, therefore, the order of loading is important. For elastic conditions to hold, extreme fibre stresses should be kept below values in Section 2.2.10, and slip at the interface between the concrete and steel should be negligible.

The elastic section properties can be evaluated from the transformed section, as in Fig. 2.15. The area of concrete within the profile depth is ignored (this is conservative where the decking troughs lie parallel to the beam). The concrete can usually be assumed to be uncracked under positive (sagging) moment.

The elastic neutral axis depth, y_e (below the upper surface of the slab), may be determined from the formula

$$y_e = \frac{\dfrac{D_s - D_p}{2} + \alpha_e r\left(\dfrac{D}{2} + D_s\right)}{(1 + \alpha_e r)}$$

where $r = A/[(D_s - D_p)B_e]$
D_s is the slab depth
D_p is the profile deck height
A is the cross-sectional area of the beam of depth D
r represents the relative proportions of steel and concrete.

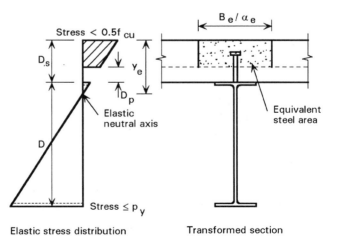

Elastic stress distribution   Transformed section

Figure 2.15   Elastic behaviour of composite beam.

The second moment of area of the uncracked composite section is

$$I_c = \frac{A(D + D_s + D_p)^2}{4(1 + \alpha_e r)} + \frac{B_e(D_s + D_p)^3}{12\alpha_e} + I_s$$

where I_s is the second moment of area of the steel section.

The section modulus for the steel in tension is

$$Z_t = \frac{I_c}{D + D_s - y_e}$$

and for concrete in compression is

$$Z_c = \frac{I_c \alpha_e}{y_e}$$

The composite stiffness can be 3 to 5 times, and the section modulus 1.5 to 2.5 times that of the I section alone.

2.2.10 *Serviceability criteria*

Stresses in simply supported composite beams are limited to p_y in the bottom fibres of the steel section and $0.5 f_{cu}$ in the concrete slab at the serviceability limit state. No account is taken of the effect of slip on these stresses nor forces in the shear connectors at the serviceability limit state.

2.2.11 *Modular ratios*

The ratio of the elastic moduli of steel and concrete (or modular ratio α_e) depends on the type of concrete, the duration of load and the relative humidity of the environment. This is because of the effect of creep of concrete. The initial elastic modulus of lightweight concrete (density $> 1750 \, \text{kg/m}^2$) is lower than that of normal weight concrete, but the creep factor under sustained loading is proportionately smaller.

Long-term and short-term modular ratios are presented in Table 2.1. Imposed loads on floors in office buildings should be assumed to comprise two-thirds short-term and one-third long-term loading, leading to 'average' modular ratios of 10–15 for normal-weight and lightweight concrete respectively. Storage loads and other largely permanent loads should be treated as long term.

Table 2.1 Modular ratio, α_e.

Type of concrete	Modular ratio for short-term loading	Modular ratio for long-term loading
Normal weight	6	18
Lightweight	10	25

2.2.12 *Deflections*

Deflection limits for beams are specified in BS 5950: Part 1.[4] Composite beams are, by their nature, shallower than non-composite beams and often are used in structures where long spans would otherwise not be economic. As spans increase, so traditional deflection limits based on a proportion of the beam span may not be appropriate. The absolute deflection may also be important and pre-cambering may need to be considered for beams longer than 10 m.

Elastic section properties, as described in Section 2.2.9, are used in establishing the deflection of composite beams. Uncracked section properties are considered to be appropriate for the calculation of deflections. In most cases, no account is taken of the benefits of continuity. The appropriate modular ratio is calculated as above, but it is usually found that the section properties are relatively insensitive to the precise value of modular ratio. The effective breadth of the slab is the same as that used in evaluating ultimate strength.

The deflection of a simple composite beam subject to unfactored imposed load is modified by the effects of partial shear connection. The modified deflection δ'_c (for $K \leq 1$) is obtained from the following formulae

$$\delta'_c = \delta_c + 0.5(1 - K)(\delta_0 - \delta_c) \qquad \text{for propped beams}$$

$$\delta'_c = \delta_c + 0.3(1 - K)(\delta_0 - \delta_c) \qquad \text{for unpropped beams}$$

where δ_c and δ_0 are the deflections of the composite and steel beam respectively at the appropriate serviceability load

K is the degree of shear connection used in determining the plastic moment capacity of the beam (see Section 2.2.6).

The difference between the coefficients in these two formulae arises from the different shear connector forces and hence slip at serviceability loads in the two cases.

In cases where the absolute deflection of the underside of the steel beam is important, the deflection of the steel beam resulting from its self-weight and that of the slab should be added to the imposed load deflection of the composite beam. In propped construction, this load is applied to the composite section on removal of the props.

2.2.13 *Dynamic sensitivity*

Shallower beams imply greater flexibility and although the in-service performance of composite beams and floors is good, designers may be concerned about the susceptibility of the structure to vibrations induced by the activities within the building. The parameter commonly associated with this effect is the natural frequency of the floor or beams.

A lower limit of 4 Hz (cycles per second) is a commonly accepted lower bound to the natural frequency of each floor beam. The mass of the floor is taken as its self-weight and that of ceiling and finishes, and 10% of the imposed loading. Partitions increase the damping of the structure and are not included.

The natural frequency of the floor or beam may be determined from the approximate formula $f = 18/\sqrt{\delta}$, where δ is the instantaneous deflection resulting from the self-weight of the floor (including the above additional loads). A 10% reduction in deflection may be made to account for the increased dynamic stiffness of the composite beam.

2.2.14 *Fire condition*

The fire resistance of composite beams is assessed in the same manner as for non-composite beams. According to BS 5950: Part 8[2] the limiting temperature of the steel section can be established, and this is used in determining the required thickness of fire protection. It is traditional practice to seal the voids created by the deck above the top flange of the beam, although this may not be necessary for dovetailed profiles, as in Fig. 2.3.

Chapter 3
Slimflor Construction

3.1 Background information

In the last decade, the Nordic countries, Sweden in particular, have developed a method for constructing floors using beams colloquially known as 'top hats'. These beams are fabricated out of plates to form the shape of a top hat (see Fig. 3.1). Adopting this method of construction has dramatically increased the market share for steel framed multi-storey buildings. In view of this success, British Steel plc decided to send a team of structural engineers including the author to Sweden to investigate the merits of this system. The findings of that investigation supplemented by further research into Slimflor construction have been used to develop the Slimflor beam, (see Fig. 3.2).

This method of construction is marketed by British Steel plc under the trademark 'Slimflor'. The Slimflor beam concept was introduced to the UK in November 1991. Since the introduction of this form of construction a high level of enthusiasm has been shown by specifiers. In addition, the Slimflor concept has attracted worldwide attention.

Initially, the Slimflor method of construction was used in conjunction with precast concrete floor units, and later the development of profiled steel deep decking. Precision Metal Forming (PMF) developed the deep deck profile (CF210) which has a depth of 210 mm and is trapezoidal in cross-section (see Section 3.2). The steel deck reduces the dead weight of the floor and improves the 'robustness' of the construction. Also, it lends itself to the accommodation of services within the slab depth between the ribs of the deck. Holes constructed through the web of the beam enable these services to pass through the

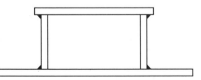

Figure 3.1   Top-hat beam.

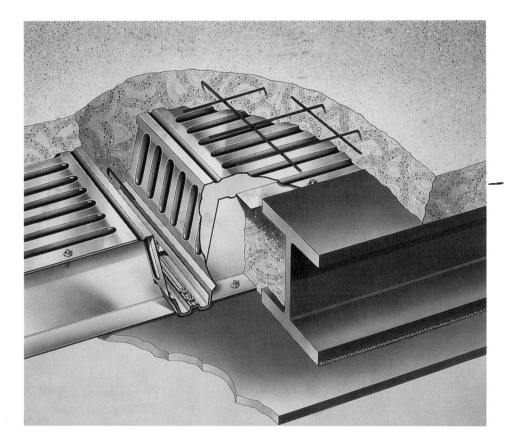

Figure 3.2   Basic components of the Slimflor system using deep profiled steel decking.

floor (see Fig. 3.8). Although precast concrete floor units can be used to form the floor slab, this section will concentrate mainly on the use of deep profile decking.

This chapter covers the basic design principles for Slimflor construction. (For further information see References 5, 6, 7 and 8.)

3.2 Slimdek construction

Further research and development into the Slimflor approach has lead to the Slimdek® method of construction. Slimdek is a British Steel plc trademark.

Slimdek is the collective term for the following methods of Slimflor construction:

(1)   Slimflor Beam (SFB)
(2)   Asymmetric Slimflor Beam (ASB)

(3)   RHS Slimflor Edge Beam (RHSFB)
(4)   A new deep profiled steel deck (SD225).

The three basic forms of construction above ((1), (2) and (3)) have been designed to be used with the new deep deck profile (SD225). This new deep deck profile replaces the existing CF210 deck.

The different forms of Slimdek construction are fully explained in this section. To emphasise the method of analysis, worked examples and derivation of formulae are provided in Appendix B.

3.3   The general benefits of using Slimdek construction

- Floors have a flat soffit which offers unhindered passage for the services.
- The overall floor construction depth is reduced. This has an influence on the cladding costs i.e. the overall height of the building is reduced.
- Slimflor construction is comparable to other 'fastrack' methods for speed of construction. In fact, the Slimflor approach is now seen as a logical progression to the shallow profiled decking method of composite construction as illustrated in Chapter 2.
- It improves the fire resistance of the section. The concrete that surrounds the beam partially insulates the section, giving in the order of 60 minutes fire resistance. This can therefore eliminate the need for additional fire protection.
- The concrete that surrounds the beam produces an increase in the second moment of area of the section. This enhancement is helpful in reducing deflections.
- In the case of local element instability, the concrete will improve the load-carrying characteristic of the beam. For the future, this could prove an asset for continuous construction.

3.4   The Slimflor beam

The floor components (i.e. pre-cast units or steel decking) span between beams and rest on the bottom flange plate. Figure 3.3 shows the basic steel components of the system.

The Slimflor beam shown in Fig. 3.3 has the following advantages:

- It uses standard steel sections.
- The beam is easy to fabricate with full depth end plate connections. Only two fillet welds are required to attach the longitudinal plate which can be automatically welded without turning the section.

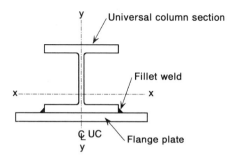

Figure 3.3   The basic steel components of the Slimflor beam.

● The system provides relatively long spans with minimum construction depths.
● No internal voids for sound or heat transfer in fire are created. This reduces the amount of fire protection.

3.4.1   *Slimflor beam combined with the composite slab*

In the deep deck system (see Section 3.7 for further information about the deck), the deck sits on the bottom plate and spans between the beams. It acts as permanent formwork to the in situ concrete slab and also develops composite action with the concrete.

A special end diaphragm to the decking is used so that the concrete fully surrounds the steel section except for the bottom surface of the plate and in some cases the upper surface of the top flange. The insulating effect of the concrete provides the beam with 60 minutes fire resistance. Higher periods of fire resistance may be achieved with additional fire protection to the plate. The soffit of the deck is left exposed and fire resistance of the slab is achieved by the provision of additional reinforcing bars in the ribs.

The Slimflor beams may be designed economically within the constraints of the slab depth for spans of 6–9 m. Two forms of construction are considered:

● *Type A*: Non-composite beam (utilising the concrete encasement only for increased stiffness and fire resistance).
● *Type B*: Composite beam with additional shear connectors and achieving composite action with the slab.

These two forms of construction are illustrated in Fig. 3.4. In principle, the beams are designed to be unpropped during construction, although there may be circumstances where propping is used to

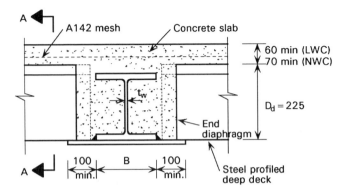

Cross-section through non-composite 'Slimflor' beam - Type A

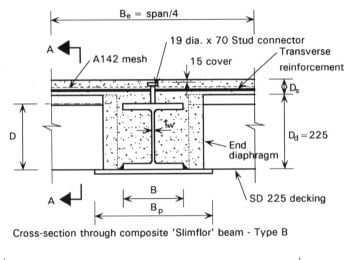

Cross-section through composite 'Slimflor' beam - Type B

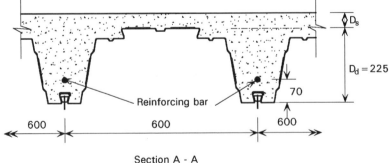

Section A - A

Figure 3.4   Construction details for Type A and Type B Slimflor construction.

provide the minimum steel beam size for a given span. Grade S355 steel and lightweight concrete are the preferred materials. However, where serviceability criteria control the design, use of grade S275 is more cost effective.

3.4.2 *Construction details*

The following construction details and notes are typical of recent good practice, and relate to the deep decking form of construction.

3.4.3 *Typical details for internal beams*

Figures 3.5 and 3.6 show typical cross-sections for various forms of internal Slimflor beams. The deck, is generally orientated perpendicular to the beam and rests on the extended bottom flange plate.

Figure 3.5(a) shows the non-composite form of construction (Type A). This type of flooring arrangement would be suitable for office construction for beam spans of 6–8 m and imposed loads of up to 4–5 kN/m^2.

The form of construction shown in Fig. 3.5(b) (Type A – modified) takes account of the use of a deeper beam projecting above the slab and can only be used where there is a raised access floor. The purpose of the step in the concrete slab is to reduce the overall depth of in situ concrete which in turn will reduce the dead weight of the floor construction and hence provide a more economical design leading to spans of up to 10 m. The reinforcing bars can be cranked over the Universal Column (UC) top flange to allow for code requirements with respect to robustness. Alternatively, the bars in the trough can be continued through holes in the beam web.

Generally, composite construction (see Fig. 3.6) is used where the building height exceeds four storeys and the imposed loads are relatively high (5 kN/m^2 and above). The form of construction makes it easier to cater for robustness and to transfer lateral forces to the core areas.

Composite action is achieved using 19 mm diameter studs which are shop welded to the top flange. For a stud height of 70 mm (after welding) and 15 mm cover, the depth of concrete above the UC top flange is 85 mm.

For this form of construction to be effective, the UC section should be confined to the 203 or 254 UC sections otherwise the concrete depth above the deck depth (D_s) could limit the spanning capabilities of the deck.

3.4.4 *Tie members*

Tie members may be permanent or provided for temporary erection purposes only. The inverted tee section shown in Fig. 3.7 may be considered as a tie member but should be designed for a partial

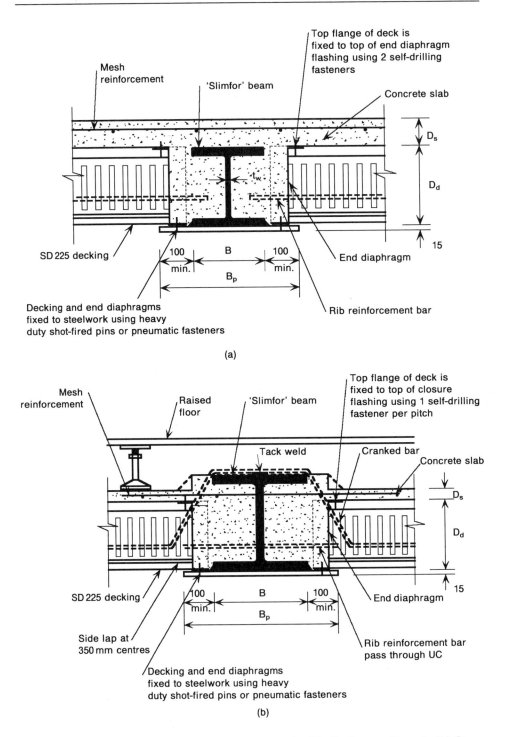

Figure 3.5 (a) Cross-section through non-composite Slimflor beam – Type A. (b) Cross-section through non-composite Slimflor beam – Type A (modified).

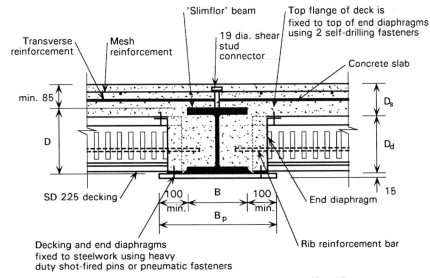

Figure 3.6   Cross-section through composite Slimflor beam – Type B.

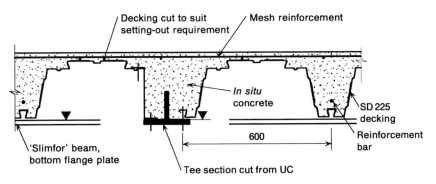

Figure 3.7   Typical cross-section through tie member.

uniformly distributed load. This is to allow for a small proportion of the concrete slab to be supported by the tee section in the construction stage. An allowance of say $3\,m^2$ of slab load should be sufficient to cater for the design of the tee section. Particular care should be taken when considering the vertical deflection of the tee section.

Figure 3.7 incorporates a setting-out tolerance for the fixing of the deck. The two upper surfaces shown with a solid triangle are to remain at the same level. Where the centre lines of the trough and columns coincide, it may be possible to use a tie member that is embedded into the slab. Temporary ties may be used for buildings of four storeys or less, provided the tie member satisfies the requirements for stability during construction. Figures 3.9 and 3.10 highlight the use of a temporary tie.

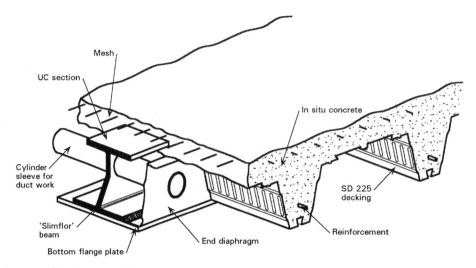

Mesh

UC section

In situ concrete

Cylinder
sleeve for
duct work

SD 225
decking

'Slimflor'
beam

End diaphragm

Reinforcement

Bottom flange plate

Figure 3.8   Cut-away view showing installation of service ducts.

3.4.5 *Holes in beam webs*

The cut-away view in Fig. 3.8 shows that the deck shape is ideal for passing minor services through the slab depth between the ribs of the deck. The cylindrical sleeve passes through the web of the UC and the end diaphragm. The in situ concrete surrounds the sleeve and encases the Slimflor beam. The end diaphragms retain the concrete that surrounds the beam.

3.5 Typical Slimflor project

The photographs shown in Figs 3.9, 3.10 and 3.11 are a good example of Slimflor construction. The building was designed by the Laing Technology Group and John Laing Construction were the main contractors.

Figure 3.9 shows the deep deck being fixed to the end diaphragms. The end diaphragms are connected to the bottom flange plate of the Slimflor beam. In the background, the deck is in a bundle ready to be passed over the end diaphragms. The deck is then fixed to the bottom flange plate by means of shot fired pins.

This view highlights how the diaphragms automatically align the deck during construction. Figure 3.10 shows the completed floor construction. Figure 3.11 shows the first floor decked-out and ready to cast the concrete slab.

Note that the original deep decking profile (CF210) was used on this project.

Figure 3.9   Fixing of the end diaphragms.

Figure 3.10   Deck construction completed – viewed from below.

Figure 3.11   General view of the floor prior to mesh placement and concreting.

3.6   Design of Slimflor beams using deep decking

3.6.1   *Basis of design*

There are two methods of construction:

- *Type A*: Non-composite beam (but using a composite slab).
- *Type B*: Composite beam and slab.

The design procedures are based on the use of BS 5950: Part 3: Section 3.1[3] and Part 1[4] for unpropped simply supported beams with uniformly distributed loading.

The steel cross-sections will be limited to plastic (Class 1) or compact (Class 2) sections. Semi-compact (Class 3) sections can be used but they limit the design to the elastic moment capacity which complicates the design procedures and, more importantly, results in uneconomical use of the steel. In addition, BS 5950: Part 3: Section 3.1 recommends plastic design of cross-sections.

Slimflor beams may be subject to out-of-balance loading either during construction or in service.

The main design assumptions are:

(1) Unpropped simply supported beams are subject to uniformly distributed loading.
(2) Only plastic or compact cross-sections are used.
(3) Plastic analysis of the cross-section is based on rectangular stress blocks.
(4) Moments and forces are determined using factored loads.
(5) Serviceability checks are determined using unfactored loads. To ensure that irreversible deformation (under normal service loads) does not occur in the steel, the extreme fibre stress is limited to p_y. The in situ concrete stress is likewise limited to $0.5f_{cu}$. However, serviceability stresses are unlikely to be a design criterion for this method of construction.
(6) Deflections of beams are limited to span/360 under imposed loads and span/200 under total load. These limitations for deflection apply to buildings of general usage. In addition, it is a requirement of BS 5950: Part 1 that due allowance should be made where deflections under serviceability loads could impair the strength or efficiency of the structure or its components or cause damage to the finishings.

3.6.2 Construction stage

In the construction stage, end diaphragms are connected to the bottom flange plate. These end diaphragms have the same profile shape as the deck, which effectively provides a closure for the concrete. The steel deck is placed over these diaphragms and connected using shot fired pins to the projecting bottom flange plate of the Slimflor beam. The in situ concrete is then poured or pumped onto the deck and levelled to a final finish. This operation may produce out-of-balance loads on the beam, the worst case being when the deck on one side of the beam is fully loaded. To account for these torsional effects a simplified approach may be used in which equal and opposite horizontal forces are applied to the flanges of the beam. A construction load of $0.5\,\text{kN/m}^2$ in addition to the self-weight of the slab and beam **must** be allowed for when considering the design of the beam.

3.6.3 *Moment resistance including lateral torsional buckling effects*

For the non-composite beam (the basic components and member axes are shown in Fig. 3.12), lateral torsional buckling (LTB) has to be considered for the construction stage unless some device is used to restrain the compression flange. Where the compression flange is considered as restrained, the full moment capacity can be developed.

The design procedures for LTB of singly symmetric sections are defined in BS 5950: Part 1.

Buckling resistance moment, M_b

$$M_b = S_x p_b$$

where S_x is the plastic modulus of the section about the x–x axis. This can be calculated from first principles using the stress block method.

p_b is the bending strength of the section, which is related to the equivalent slenderness, λ_{LT}, the design strength of the material, p_y, and member type which in this case is classed as 'rolled'. Once λ_{LT} has been established, p_y and Table 11 (BS 5950: Part 1) are used to evaluate p_b. Table 12 (bending strength p_b for welded sections) is not used because the welding takes place in the tension zone and the upper compression flange is part of the rolled section.

The equivalent slenderness, λ_{LT}

$$\lambda_{LT} = nuv\lambda$$

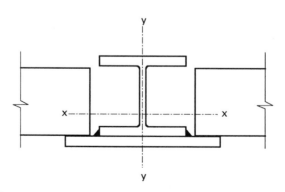

Figure 3.12   Basic components and member axes of non-composite beam.

where n, u and v are defined as follows:

n is the slenderness correction factor (from Table 13 (BS 5950: Part 1), $n = 1.0$, conservatively).
u is the buckling parameter. For flanged sections symmetrical about the minor axis

$$u = \left(\frac{4S_x^2 \gamma}{A^2 h_s^2}\right)^{1/4}$$

where S_x is as described above
A is the cross-sectional area

$$h_s \approx D - \frac{T}{2}$$

$$\gamma = 1 - \frac{I_y}{I_x}$$

I_y is the second moment of area about the y–y axis
I_x is the second moment of area about the x–x axis.

Generally, I_x is the major axis of the section but when using this form of construction this is not always the case. With the plate being welded to the bottom flange of the UC it is possible for $I_y > I_x$. In this instance, as I_y tends to I_x the γ factor becomes zero. When this occurs, the buckling parameter, u, also becomes zero. Hence, λ_{LT} is equal to zero which nullifies the requirements for LTB and failure by LTB does not occur.

This variable v is the slenderness factor. Two methods exist for obtaining the value of v, either by using Table 14 (BS 5950: Part 1)[4] or alternatively the expression given for v in Appendix B of BS 5950: Part 1. The following explains both methods:

(1)   Using Table 14 (BS 5950: Part 1).
       To use Table 14, three parameters require evaluation:

       (a)   Torsional index, x
       (b)   Flange ratio, N
       (c)   Minor axis slenderness, λ

       (a)   Torsional index, x

$$x = 0.566 \, h_s \left(\frac{A}{J}\right)^{1/2}$$

where h_s and A are as previously defined

$$J = J_{uc} + \tfrac{1}{3}(t_p^3 B_p)$$

t_p is the flange plate thickness

B_p is the flange plate width.

The torsion constant, J_{uc}, can be obtained from published tables.

(b)   Flange ratio, N

$$N = \frac{I_{cf}}{I_{cf} + I_{tf}}$$

where I_{cf} is the second moment of area of the compression flange about the y–y axis

$$I_{cf} = \frac{TB^3}{12}$$

I_{tf} is the second moment of area of the tension flange about the y–y axis

$$= \frac{TB^3}{12} + \frac{t_p B_p^3}{12}$$

(c)   Minor axis slenderness, λ

$$\lambda = \frac{L_E}{r_y}$$

where L_E is the effective length of the member which may be taken as $L_E = 1.0L_1$. The loads are applied through the tension flange which represents a stabilising condition. However, no account is taken of this beneficial effect. The radius of gyration, r_y, is taken about the minor axis of the member.

(2)   Using the expression given for v in Appendix B of BS 5950: Part 1.

$$v = \left\{ \left[4N(1 - N) + \frac{1}{20}\left(\frac{\lambda}{x}\right)^2 + \psi^2 \right]^{1/2} + \psi \right\}^{-1/2}$$

The only parameter to be determined is the monosymmetry index, ψ

$$\psi = 0.8\,(2N - 1) \quad \text{for } N > 0.5$$

$$\psi = 1.0\,(2N - 1) \quad \text{for } N < 0.5$$

The above information enables the designer to obtain the buckling resistance moment, M_b.

3.6.4 Non-composite beam (Type A)

Design assumptions

(1) The Slimflor beam is considered to be laterally unrestrained for the construction stage.
(2) The Slimflor beam is laterally restrained for the imposed loading condition.
(3) Out-of-balance loads do not need to be considered for the imposed loading provided that adequate continuity reinforcement is allowed for in the design.

3.6.5 Section classification

The cross-section (the basic section dimensions are shown in Fig. 3.13) will be limited to plastic or compact UC sections.
    The limiting width-to-thickness ratios for a compact section are:

Flange $\dfrac{b}{T} \leq 8.5\epsilon$

Web (generally) $\dfrac{d}{t_w} \leq \dfrac{98\epsilon}{\alpha}$

where $\alpha = \dfrac{2y_p}{d}$

$\epsilon = 1.0$ for grade S275 steel
$\epsilon = 0.88$ for grade S355 steel.

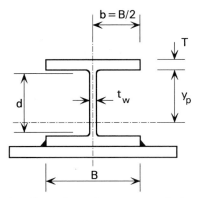

Figure 3.13   Section dimensions of non-composite beam.

If $\alpha \approx 2$ (y_p tends to d), then the section should be taken as having compression throughout. The limiting classification check then becomes

$$\frac{d}{t_w} \leq 39\epsilon$$

All UC sections of grades S275 and S355 satisfy this criterion. However, this may not apply to other shaped sections.

3.6.6 *Biaxial stress effects in the flange plate (Figs 3.14, 3.15 and 3.16)*

Biaxial stresses have to be considered as a direct result of the way the loads are applied to the flange plate. It is conservatively assumed that the shear force from the slab is applied to the flange plate at the centre of bearing of the steel deck. This assumption ignores the

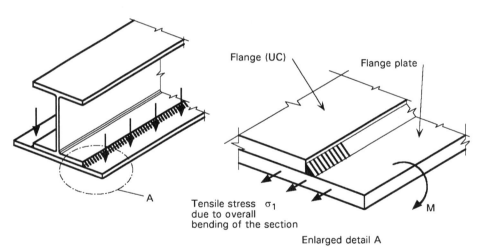

Figure 3.14 How loads are applied to the bottom flange plate.

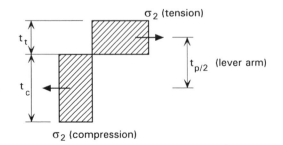

Figure 3.15 Plastic distribution through the flange plate.

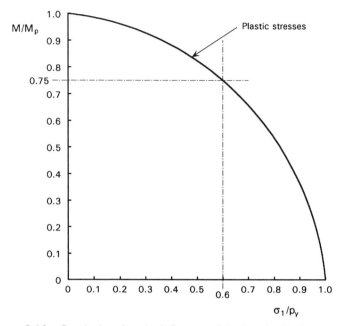

Figure 3.16   Graph showing the influence of the longitudinal stress on the transverse bending stress for the plastic.

transfer of shear directly through the concrete ribs which could greatly reduce this action. However, the following approach is adopted in order to offer a 'fail-safe' design independent of concrete properties. This approach also ensures independent load transfer during construction. The plate is subject to longitudinal and transverse effects. The longitudinal stress due to overall bending of the section, σ_1, has an influence in reducing the resistance of the plate when also subject to a transverse bending stress, σ_2. This is irrespective of whether the stresses are plastic or elastic.

Plastic analysis
The effective transverse yield stress that may be resisted is reduced from p_y to σ_2 which according to von Mises yield criterion is given by

$$[\sigma_2^2 - \sigma_1\sigma_2 + \sigma_1^2]^{1/2} = p_y$$

or

$$\sigma_2 = \frac{\sigma_1 \pm (4p_y^2 - 3\sigma_1^2)^{1/2}}{2}$$

Taking a positive sign for compression

$$\sigma_2 \text{ (compression)} = \frac{\sigma_1 + (4p_y^2 - 3\sigma_1^2)^{1/2}}{2}$$

$$\sigma_2 \text{ (tension)} = \frac{\sigma_1 + (4p_y^2 - 3\sigma_1^2)^{1/2}}{2}$$

Using the above equations it can be shown that

$$\frac{M}{M_p} = \frac{c^2 - \sigma_1^2}{2cp_y}$$

where $c = (4p_y^2 - 3\sigma_1^2)^{1/2}$

M is the maximum transverse moment applied to the plate

M_p is the moment capacity of the plate $= \dfrac{T^2 p_y}{4}$.

The graph shown in Fig. 3.16 plots (M/M_p) on the vertical axis and (σ_1/p_y) on the horizontal axis. This direct relationship between these two ratios gives engineers a visual indication of the extent to which the longitudinal stress, σ_1, will influence the transverse bending stress, σ_2.

For example, say, $\sigma_1 = 0.6\,p_y$ due to overall bending, then the plate transverse moment capacity

$$M \approx 0.75\,M_p$$

$$= \frac{3}{4} \left(\frac{T^2 p_y}{4} \right)$$

$$= \frac{3}{16} (T^2 p_y) \text{ per unit length of beam}$$

3.6.7 Construction stage effects

In the construction stage it is not always possible to control the way the concrete is poured onto the steel deck. For the construction stage it would be prudent to assume that an internal bay is fully loaded with the adjacent bays unloaded (see Fig. 3.17).

There are two methods of analysis of the treatment of out-of-balance loads which are:

(1) Simplified approach to torsion.
(2) Rigorous method of analysis.

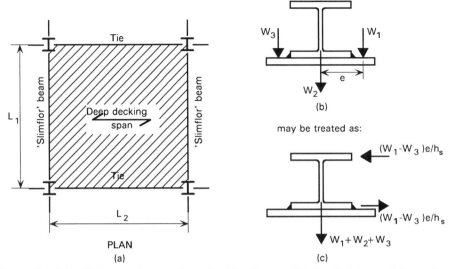

Figure 3.17 Fully loaded internal bay. (a) Plan layout. (b) Applied loads. (c) Torsional effects.

The latter method of analysis has been dealt with in the SCI publication *Design of members subject to combined bending and torsion*.[9] This method of analysis has been adopted for the Slimflor computer software. Also, this method of analysis has been used in the worked examples (see Appendix B).

Figure 3.17(a) shows a plan layout for an internal bay with typical cross-sections through the floor beam. The hatched area represents a fully loaded bay and the enlarged cross-section shows the applied loads (Fig. 3.17(b)). Torsional effects may be treated for simplicity in terms of warping of the cross-section and ignoring the pure torsional resistance. This may be treated by considering equal and opposite transverse forces in the flange (Fig. 3.17(c)) in equilibrium with the applied torsion.

It should be pointed out that the simplified approach is conservative (25% in some cases) when compared to the rigorous method of analysis.

An additional criterion is that the connections between the Slimflor beam and the columns should be capable of resisting the total torsional moment transferred to the beam.

The method adopted is to check the section using the simplified approach and satisfy the following 'unity factor' condition

$$\frac{M_x}{M_b} + \frac{M_y}{M_{cy}} \leq 1.0$$

where M_x is the applied moment about the x–x axis
M_b is the buckling resistance moment about the x–x axis (determined using BS 5950: Part 1 and Section 3.6.3)
M_y is the applied transverse moment to the top flange of the UC about the y–y axis. This is treated by considering the torsional moment as two opposing forces in the flanges.
M_{cy} is the moment resistance of the top flange (UC) about the y–y axis $= \dfrac{TB^2 p_y}{4}$.

The formulae required to evaluate the above expression are:

Applied moment, M_x

Factored loading

$$W_1 = [(w_d + w_r + w_{cc})\gamma_{fd} + (w_c\gamma_{fc})]\left(\frac{L_2 L_1}{2}\right)$$

$$W_2 = w_s\gamma_{fd}L_1 L_2$$

$$W_3 = [(w_d + w_r)\gamma_{fd}]\left(\frac{L_2 L_1}{2}\right)$$

where w_d is the weight of steel deck etc.
w_r is the weight of reinforcement
w_{cc} is the weight of in situ concrete
w_c is the construction load
w_s is the self-weight of steel
γ_{fd} is the load factor for dead loads (1.4)
γ_{fc} is the load factor for imposed loads (1.6).

Applied moment about the x–x axis, M_x

$$M_x = (W_1 - W_2 + W_3)\frac{L_1}{8}$$

Transverse moment, M_y

The equivalent horizontal force, F, due to torsional effects on the section (see Fig. 3.17) is

$$F = (W_1 - W_3)\frac{e}{h_s}$$

where $h_s \approx D - \dfrac{T}{2}$

e is the eccentricity of load W_1 measured from the centre line of the UC section.

$$M_y = \frac{FL_1}{8}$$

3.6.8 Vertical shear resistance

The vertical shear resistance, P_v, of the steel member is determined using BS 5950: Part 1.

$$P_v = 0.6 p_y A_v$$

where A_v is the shear area taken as $t_w D$.

Vertical shear can influence the moment resistance of the beam. This occurs where high shear and moment co-exist at the same position within the span. Simply supported beams with one or two point loads are good examples of cases where this occurs.

3.6.9 Moment resistance of non-composite beam (laterally restrained)

The concrete that surrounds the beam is primarily used for stiffness purposes and to provide lateral restraint to the member at the ultimate limit state. Enhancement of the moment resistance of the beam by combining the surrounding concrete with the steel member is difficult to justify unless additional shear connection is provided. It is for these reasons that the composite action is neglected at the ultimate limit state, thereby simplifying the design procedures. The method of deriving the equation for moment capacity (steel member only) using plastic analysis for the design of the cross-section is described below.

Figure 3.18 shows the position of the plastic neutral axis (pna) in the web, at a distance y_p below the centre line of the UC. To analyse the moment capacity from this diagram would involve some tedious calculations. In order to simplify the calculation, Fig. 3.18 also illustrates a standard method of rearranging the rectangular stress blocks.

The value of y_p can be found by equating the tensile resistance of the plate, R_p, to the increased compressive zone in the web. Hence

$$y_p = \frac{R_p}{2 p_y t_w}$$

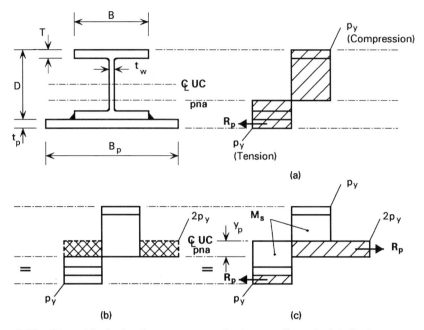

Figure 3.18 Stress blocks for the non-composite beam – Type A. (a) Basic rectangular stress blocks. (b) Addition of self-equilibrating stress blocks (shown by cross-hatching). (c) Simplified plastic analysis.

Moments are now taken about the centre line of the UC to find M_c, such that

$$M_c = M_s + R_p \left(\frac{D}{2} + \frac{t_p}{2} \right) - R_p \frac{y_p}{2}$$

$$= M_s + \frac{R_p}{2}(D + t_p) - \frac{R_p^2}{4p_y t_w}$$

where $M_s = S_x p_y$
$R_p = A_p p_y$
A_p is the area of flange plate $(B_p t_p)$.

3.6.10 *Composite beam (Type B)*

The main design assumptions are as for the non-composite beam (see Section 3.6.4).

Effective breadth of compression flange
The effective breadth of slabs, B_e, will vary for ultimate and serviceability limit states but a compromise value of span/4 for

internal beams has been made in BS 5950: Part 3: Section 3.1. The value of B_e should not exceed the distance between beam centres.

Modular ratio (see Section 2.2.11)
The modular ratio is used for serviceability calculations and is the ratio of the elastic moduli of steel and concrete. For buildings of normal usage, the modular ratio should be assumed to be in the ratio of 2/3 short term and 1/3 long term, reflecting the influence of creep of concrete. This gives values of 10 and 15 for normal and lightweight concrete respectively for use in imposed load calculations.

3.6.11 *Moment capacity of composite beam*

The moment capacity of composite Slimflor beams is dependent on the degree of shear connection between the concrete and steel beam. The relevant design cases are as follows.

Full shear connection
Full shear connection occurs where the number of shear connectors provided is at least equal to that required to develop the full resistance of the concrete or the steel member. This will generate the maximum moment capacity of the cross-section. In the majority of cases, the concrete resistance, R_c (see Fig. 3.19), is less than the resistance of the steel member, $(R_s + R_p)$

where $R_c = 0.45 f_{cu} B_e D_s$ if $D_d \geq (D - T)$
or $R_c = 0.45 f_{cu} B_e (D_s + D_d - D + T)$ if $D_d < (D - T)$
         $R_s = A p_y$
         $R_p = A_p p_y$
         D_s is the depth of the slab above the top of the profiled deck.

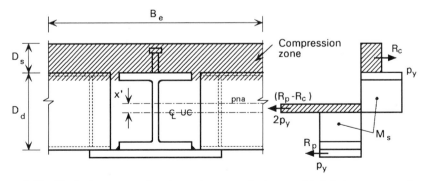

Figure 3.19 Typical cross-section through composite beam with full shear connection.

Moment capacity for full shear connection, M_c

$$M_c = M_s + R_c\left(D_d + \frac{D_s}{2} - \frac{D}{2}\right) + \frac{R_p}{2}(D + t_p) - \frac{(R_c - R_p)^2}{4p_y t_w}$$

where $M_s = S_x p_y$
S_x is the plastic section modulus of the UC
D_d is the depth of steel deck.

Partial shear connection

Partial shear connection is used when there is an excess of moment capacity compared to the applied factored moment. In this situation, BS 5950: Part 3: Section 3.1 allows a reduction (to a maximum of 40%) in the number of shear connectors needed for full shear connection.

The principle is that the number of shear connectors may be reduced so that the force transferred, R_q (see Fig. 3.20), is sufficient to provide the required moment capacity. In this case, the force in the slab is R_q (not R_c) and the moment capacity may be determined as follows.

Moment capacity for partial shear connection, M_c

$$M_c = M_s + R_q\left(D_d + D_s - \frac{D}{2} - \frac{y'}{2}\right)$$

$$+ \frac{R_p}{2}(D + t_p) - \frac{(R_q - R_p)^2}{4p_y t_w}$$

where

$$y' = \frac{R_q}{0.45 f_{cu} B_e}$$

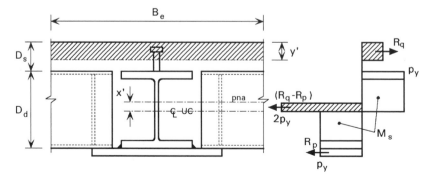

Figure 3.20   Typical cross-section through composite slab with partial shear connection.

R_q is given by the number of shear connectors in the half span multiplied by the strength of each shear connector, determined from Table 5 of BS 5950: Part 3, reduced by a factor of 0.8 (and further by 0.9 for lightweight concrete) – see the following section on shear connector design.

Note: When $y' = D_s$, then $R_q = R_c$. Assume a degree of shear connection, e.g. 40% such that $R_q = 0.4R_c$. Calculate moment capacity and compare to applied factored moment.

If $M_c <$ applied factored moment, then repeat the procedure with a higher degree of shear connection until $M_c \geq$ applied factored moment.

Deflection

The reason for Slimflor construction is, as the name implies, to provide the minimum in floor depth construction. However, shallow beam depths with relatively long spans are sensitive to deflection.

The deflection limits for beams are given in BS 5950: Part 1. The imposed load limit is span/360 based on unfactored loads.

Propping
Propping of beams or decking is not generally recommended for the spans normally used in buildings. However, for long-span beams (>9 m span) or slabs (> 6 m span) a central line of props may be required.

Partial shear connection
As a consequence of the methods used for determining the moment capacity, an increase in vertical deflection will have to be allowed for under serviceability loads.

Partial shear connection design results in a greater degree of slip occurring in the shear connectors. The result of this action is an increase in vertical deflection. This is given by the following expression for unpropped construction

$$\delta = \delta_c + 0.3\left(1 - \frac{N_a}{N_p}\right)(\delta_s - \delta_c)$$

where δ_c is the the deflection of the composite beam under imposed load

δ_s is the the deflection for the steel beam acting alone under imposed load

N_a/N_p is the degree of shear connection.

Shear connector design

As already stated the most popular size of shear stud connector available is the 19 mm diameter × 100 mm high (95 mm after welding). However, to keep the in situ concrete to a minimum thickness over the top flange (UC) it is proposed to use the 19 mm diameter × 75 mm high (70 mm after welding) stud. To prevent 'burn through', the diameter of the stud should not exceed 2.5 times the thickness of the beam flange (not normally a problem with UC sections). Allowing for 15 mm cover, the slab depth over the flange becomes 85 mm (minimum depth). Any unnecessary increase in the concrete depth will obviously add further to the dead loads and produce greater construction depths.

The studs are welded to the top flange of the beam in the works prior to sending out to site. This procedure has the advantage over site welding in avoiding the weather conditions which can interfere with the welding procedures.

Shear connector spacing

The maximum longitudinal spacing between the shear connectors should not exceed 600 mm or four times the slab depth. In this instance, the slab depth is taken as the lesser of the depth of in situ concrete above the top flange (UC) or the deep decking. Figure 3.21 shows a plan of the beam top flange with the limiting dimensions for stud spacing.

The minimum longitudinal spacing is five times the stud diameter, d_s, and the minimum transverse spacing is four times the stud diameter (for studs in pairs), except when the studs are 'staggered'.

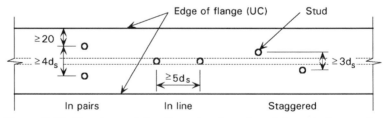

Figure 3.21 Requirements for dimensional spacing of shear stud connectors.

3.6.12 Serviceability stresses

Stresses at the serviceability limit state are calculated to ensure that under working loads no permanent deformations can occur in the steel member. The stress in the extreme fibre of the steel beam should not exceed the design strength, p_y, and the stress in the concrete flange should not exceed $0.5f_{cu}$. Note that serviceability stresses are unlikely to be a design criterion for this method of construction.

No account is taken of the effect of slip on these stresses and the associated forces on the shear connectors at the serviceability limit state.

3.6.13 *Transverse reinforcement*

Transverse reinforcement (see also Section 2.2.8) is used to ensure a smooth transfer of the longitudinal force at the ultimate limit state (via the shear connectors) into the slab without splitting the concrete. Potential shear failure planes through the slab lie on either side of the shear connectors as shown in Fig. 3.22.

The shear resistance per unit length of each plane along the beam is given by

$$v_r = 0.8 A_{sv} f_y + 0.33 \eta A_{cv} f_{cu}$$

but $v_r \le 0.8 \eta A_{cv}$

(note, the deck contribution has been omitted from the above equation for v_r)

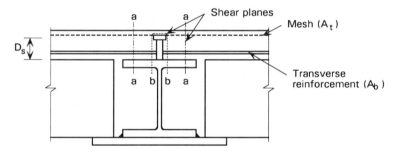

Figure 3.22   Assumed shear failure planes at the ultimate limit state.

Table 3.1   Reinforcement area and lengths of shear plane for Fig. 3.22.

Shear plane	A_{sv}	Length of shear plane	
		One plane	Two planes
a–a	$A_b + A_t$	$D_s{}^{x}$	$2D_s$
b–b	$2A_b$	—	y & z

$^{x}D_s$ is the depth of concrete above the top flange of the UC.
^{y}For one row of shear connectors, length of shear plane = $2h$ + head diameter of stud.
^{z}For two rows of shear connectors, length of shear plane = $2h + s_t$ + head diameter of stud.

where h is the height of stud, and s_t is the transverse spacing centre-to-centre of the studs.

where f_{cu} is the characteristic cube strength of the concrete in N/mm^2, but not greater than 40 N/mm^2

$\eta = 1.0$ for normal weight concrete

$\eta = 0.8$ for lightweight concrete

A_{cv} is the mean cross-sectional area, per unit length of the beam, of the concrete shear surface under consideration

A_{sv} is the cross-sectional area per unit length of the beam, of the combined top and bottom reinforcement crossing the shear surface.

The longitudinal shear force per unit length, v, to be resisted can be obtained from the spacing of the shear connectors

$$v = \frac{NQ}{s}$$

where N is the number of shear connectors in a group

Q is the shear connector value for positive moments

s is the longitudinal spacing of the shear connectors.

3.6.14 *Natural frequency of the beam*

This design check is normally considered where the spans are long but in view of these higher dead loads it would be prudent to check all beams, irrespective of the span. (See Section 2.2.13 for further information.)

3.6.15 *Edge beams*

The first consideration for the edge beam design is to determine the form of construction that is permitted. The selected form of construction for the edge beam will dictate the design concept. For example, if a downstand beam will incorporate the cladding details, a more traditional design approach can be adopted. The deep deck can rest on the top flange of the beam and the vertical loads may be assumed to act through the shear centre of the beam. This has the advantage of eliminating the torsional effects on the beam.

Edge beams which have eccentric loads will have to cater for the out-of-balance load effects in a similar manner to the procedures given for an internal beam. Alternatively, the loads can be shown to act through the shear centre of the section or the external cladding loads can balance these eccentric forces. Where a net torsional moment exists, this will have to be combined with the lateral torsional buckling of the section unless it can be shown that the compression flange of the section is otherwise restrained.

3.6.16 *Connections*

The beams are assumed to be simply supported which means that the end plates should be of a flexible nature. The use of a partial depth end plate (end plate welded to the beam web only) would provide the flexibility for the connection to be considered 'simple'. However, the stages of construction have to allow for the torsion of the beams (due to out-of-balance loads) which makes the partial depth end plate unsuitable in this instance. For these reasons, it is recommended to use a full depth welded end plate. This type of end plate detail should provide a connection suitable for withstanding beam rotations. In addition, the full depth end plate will give extra stability to the frame in the erection stage.

An end plate which is welded to the flanges and web of the beam can be considerably stiffer than a partial depth end plate. To ensure that the connection remains 'flexible' a limit is placed on the plate thickness and bolt centres. Because of the influence of torsion it is recommended that the following is adopted

End plate thickness	10 mm
Bolt cross centres	140 mm

Based on an extensive series of fire tests on Slimflor construction it is recommended that the bottom flange plate is welded to the flange of the UC with fillet welds of not less than 6 mm.

Various arrangements exist for the tie-to-column connection. In the majority of cases the tie member will have small amounts of applied load. In those circumstances the design criterion is likely to depend upon the requirements for robustness.

3.7   Slimflor beam test programme

This section covers the full-scale Slimflor beam tests carried out at City University, London, and the fire tests for the slab and beam carried out at Warrington Fire Research Centre, Warrington, Cheshire, UK.

The Slimflor beam and slab tests utilised the original CF210 deep deck profile. The testing programme for the new deep deck (SD225) is covered in Section 3.10.

3.7.1 *Fire resistance*

The three fire tests which were carried out at Warrington Fire Research Centre have been used to determine the temperature and thermal distribution through the floor components. This data was then

used to develop a computer model to predict whether the floor components require fire protection for the period of fire exposure under
consideration.

Composite slab

The first two fire tests considered the composite (steel deck and in situ
concrete) slab only. The specimens were 1.0 m square and the slab was
formed using lightweight (LWC) and normal weight (NWC) concrete
with reinforcing bars placed in the ribs of the deck. In addition, A142
mesh was placed over the deck with 35 mm cover. These tests were
required to establish the thermal characteristics of the slab.

Slimflor beam

The third test was a full-scale test which combined the slab and the
Slimflor beam (see Fig. 3.23). This beam test achieved 62 minutes fire
resistance without the use of fire protection, when subject to its design
imposed load.

3.7.2   Slimflor beam test

This test investigated the behaviour of a Slimflor beam comprising a
composite slab formed using a deep (210 mm) steel deck and acting as
formwork to the in situ concrete (see Fig. 3.24, showing the general

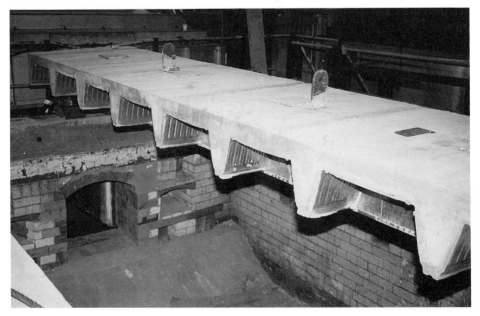

Figure 3.23   The Slimflor beam being raised from the furnace.

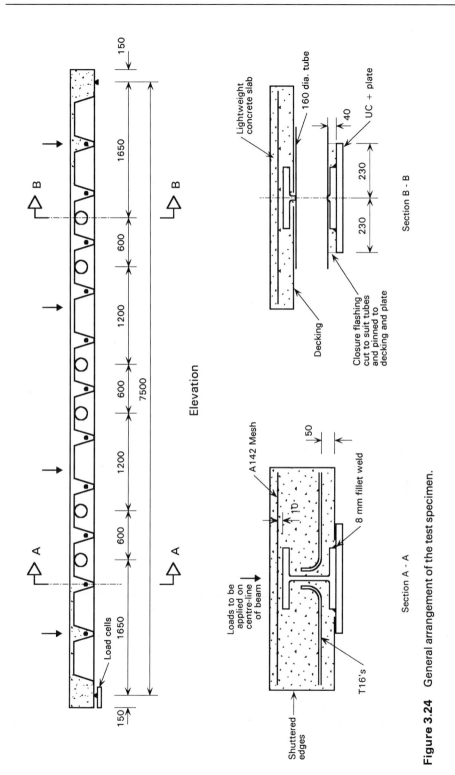

Figure 3.24   General arrangement of the test specimen.

arrangement of the test). The deck rested on a steel plate welded to the bottom flange of a UC section. The test specimen spanned 7.5 m with an overall construction depth of 300 mm. Figure 3.25 shows the specimen on its supports after striking the timber formwork. Four jacks placed on top of the beam simulated a uniformly distributed load (see Figs 3.24 and 3.26).

In addition, holes were formed in the web of the beam which permitted short lengths of cylinders to pass through the beam and

Figure 3.25   Beam on its supports after striking the formwork.

Figure 3.26   Loading arrangement (ignore raking members).

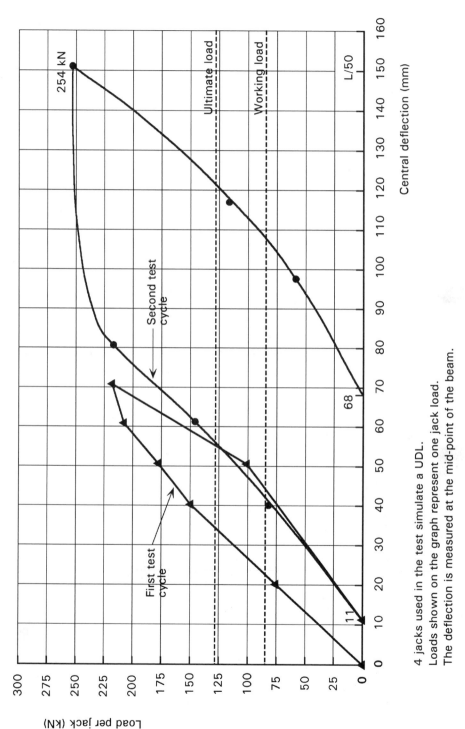

4 jacks used in the test simulate a UDL.
Loads shown on the graph represent one jack load.
The deflection is measured at the mid-point of the beam.

Figure 3.27 Slimflor beam test.

these continued to the diaphragms. This type of deep deck Slimflor beam enables minor services to be passed through the cylinder and within the decking profile and thus to be incorporated within the depth of the floor.

The test was discontinued when the total jack load had reached 1016 kN (100 tonnes) and the central deflection was 150 mm (span/50) (see Fig. 3.27). Analysis of the test results showed that the maximum moment (including self-weight) achieved in the test was 925.3 kNm. The ultimate moment of resistance for the bare steel section (using measured properties) equalled 551.7 kNm. The ratio of actual bending moment to the resistance moment based on the steel sections was 1.68.

An alternative method of illustrating the increase in capacity is to compare load intensities for a typical 6 m × 7.5 m bay. The maximum bending moment converts into a load intensity of 21.9 kN/m². Making an allowance for dead and service loads, the specimen could have supported an imposed load of 10.9 kN/m² as opposed to a typical design value of 5.0 kN/m².

The moment ratio and load intensity method, shown above, illustrates the inherent capacity of the system. However, this inherent capacity was due to the Slimflor beam acting compositely with the concrete, the measured material strengths being only slightly above design values.

This increase in moment capacity due to the composite action between the beam and concrete has been used to develop the method of analysis for the Asymmetric Slimflor Beam (ASB).

3.8   Asymmetric Slimflor Beam (ASB)

3.8.1 *Introduction*

Advances in rolling technology have turned attention to improving the economy of Slimflor construction further by developing a rolled asymmetric steel beam specifically for this application. The asymmetric section does not require welding of an additional plate and achieves optimum properties for design at the ultimate, serviceability and fire limit states. However, until the ASB range of sections is extended, it is envisaged that there will be applications where the ASB section is not suitable, i.e. longer spans are required, heavier loads, etc. For these cases the conventional Slimflor beam can be used. The ASB form of construction is illustrated in Fig. 3.28.

The additional benefits of the Asymmetric Slimflor Beam are:

- Reduced steel weight (and hence cost) in comparison with conventional Slimflor sections.

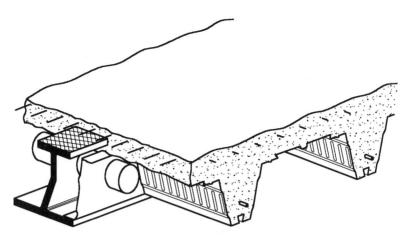

Figure 3.28 ASB construction using deep decking.

- Savings in fabrication costs.
- Readily available section with defined properties.
- Less distortion due to welding.
- Good composite action (as obtained from tests).

The ASB section is designed for use only with a deep deck composite slab because the composite properties of the ASB are used in evaluating the load–span tables for each ASB section. Initially, a limited number of ASB sections are marketed, two of which are 280 mm deep, and one 300 mm deep. All sections use S355 steel (formerly grade 50). The steel decking sits directly on the bottom flange.

Economic assessments of the use of these ASB sections have shown that the potential weight saving relative to conventional Slimflor beams is of the order of 15–25%, and the additional saving in fabrication cost is significant because it is not necessary to weld a bottom plate to the section. With these economies, ASB construction may be shown to be cheaper than conventional composite beam and slab, and reinforced concrete flat slab construction in the same medium-span range.

Coupled with the use of the ASB sections is the launch of the new deep deck profile (designated as SD225), as illustrated in Fig. 3.28.

3.8.2 *ASB section sizes*

The development of the Asymmetric Slimflor Beam (ASB) recognised the normal range of slab depths that are used in deep deck construction. For adequate fire, acoustic and structural requirements, slab depths range from 290–320 mm when using the 225 mm deep deck. The ASB section is based either on a nominally 280 mm or 300 mm deep section. The steel decking sits on the bottom flange of the section so that the depth of concrete over the ASB section is at least 30 mm.

Three ASB sections sizes in S355 steel were specifically designed to achieve the following load and span characteristics (see Table 3.2). All these sections are intended to be used *without* additional fire protection to achieve 60 minutes fire resistance.

The ASB sections are designated by their approximate depth and mass per metre (i.e. 280 ASB 100, 280 ASB 136, or 300 ASB 153). Their cross-sections are illustrated in Fig. 3.29. The bottom flange is nominally 300 mm wide and the degree of asymmetry of flange widths is 63%. The flanges of the 280 ASB 100 are 6 mm narrower so that the outstand is the same for both 280 ASB sections (which is a rolling requirement). The net projection of the bottom flange is 55 mm in all cases. Allowing for adequate end bearing, the effective span of the slab is given by the beam centres minus approximately 200 mm.

These unprotected ASB sections have been engineered to optimise all the design parameters, and to achieve good fire resistance by having a relatively thick web. The web is more effective than the exposed bottom flange in fire conditions. The saving in the welded bottom plate required in conventional Slimflor construction is much greater than the thicker web used in these ASB sections, and therefore the ASB sections are significantly lighter.

The 300 ASB section has longer spanning capabilities, and is designed so that the slab surface may be cast with 30 mm top cover, or alternatively, level with the top of the section. In the latter case,

Table 3.2 *Imposed loads and spans for ASB sections.*

Designation	Mass (kg/m)	Steel thickness (mm)		Beam span (m)	Beam spacing (m)	Imposed[a] load (kN/m^2)
		Flange	Web			
280 ASB 100	100	16	19	6	6	5.0
280 ASB 136	136	22	25	7.5	6	3.5
or				6	7.5[b]	5.0
300 ASB 153	153	24	27	7.5	7.5[b]	3.5
or				7.5	6	5.0

[a] In addition to a partition load of 1 kN/m^2.
[b] Deck is propped in this case.
  The root radius between the web and flanges is 24 mm in the 280 ASB and 27 mm in the 300 ASB.

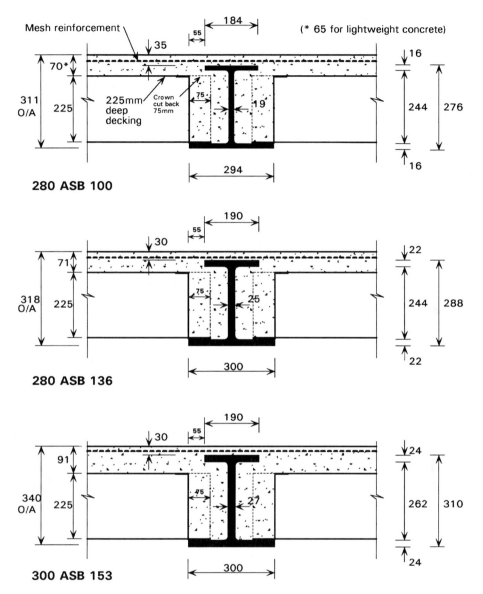

Figure 3.29 ASB section designations and minimum slab depths.

additional bars should be passed through punched holes in the web to develop the necessary tying action in the floor slab. The choice between the two methods depends on the required depth of the slab for insulation purposes in fire.

All ASB and Slimflor beams have the facility for creating circular or elongated openings of up to 160 mm diameter in the webs of the sections.

3.8.3 *Design principles*

The design of Asymmetric Slimflor Beams (ASB) follows that of conventional Slimflor beams. The ASB sections are designed to be used with the new 225 mm deep deck profile (see Section 3.10) and to act compositely with the in situ concrete encasement without use of welded shear connectors. The degree of composite action is established by tests (see Section 3.10).

The design principles for ASB sections at the ultimate, serviceability and fire limit states are described in the following sections. Additional checks are carried out at the construction stage. The design of the decking and composite slabs is reviewed in Section 3.10.

Design properties of steel sections

According to Table 6 of BS 5950: Part 1,[4] the design strength of S355 steel, p_y, is 345 N/mm² for rolled sections whose elements are of thickness greater than 16 mm. This requirement is taken to apply to *all* the ASB sections, although the flanges of the 280 ASB 100 section are 16 mm thick.

Table 3.3   Principal section properties of ASB sections.

Section properties	Units	Symbol	280 ASB 100	280 ASB 136	300 ASB 153
Mass of beam	kg/m	—	100.3	136.4	152.8
Depth of beam	cm	D	27.6	28.8	31.0
Cross-sectional area	cm²	A	127.8	173.7	194.6
Major axis properties					
Plastic section modulus	cm³	S_x	1294	1805	2159
Elastic section modulus	cm³	Z_x	995	1367	1628
Second moment of area	cm⁴	I_x	15 506	22 216	28 398
Depth to elastic neutral axis from top of section	cm	y_e	15.6	16.3	17.4
Depth to plastic neutral axis from top of section	cm	y_p	18.4	19.2	20.4
Minor axis properties					
Plastic section modulus	cm³	S_y	511	741	817
Elastic section modulus	cm³	Z_y	289	417	456
Second moment of area	cm⁴	I_y	4245	6256	6840
Radius of gyration	cm	r_y	5.76	6.00	5.92
Buckling parameters					
Torsional stiffness	cm⁴	J	160	379	513
Warping constant	cm⁶	C_w	450 943	709 504	894 772
Torsional index	—	x	13.3	10.19	9.97
Buckling parameter	—	u	0.81	0.81	0.82

By comparison, in Eurocode 3 (EC3)[10] the design strength of S355 steel is taken as $355\,N/mm^2$ (divided by a partial factor of 1.05 for designs to the UK National Application Document of EC3 Part 1.1) for steel thicknesses up to 40 mm in rolled sections.

All ASB sections are 'plastic' to BS 5950 (or 'Class 1' to EC3) and therefore their moment resistances may be calculated on plastic analysis principles.

The principal section properties and mass of the ASB sections are presented in Table 3.3. These properties also take account of the root radii between the flanges and web. Precise dimensions are given in Fig. 3.29. Tolerances on these dimensions and properties are the same as for conventional rolled sections.

Construction stage loading

The worst load cases for design of the steel beam during the concreting operation are:

- Uniformly distributed loading on one side of the beam, causing the maximum out-of-balance loads on the beam.
- Uniformly distributed loading over the entire supported area, causing the maximum bending moment on the beam.

These cases are illustrated in Fig. 3.30, where W_c is the load applied from one deck span at the construction stage. Generally, it is the out-of-balance load case which is critical because it causes transverse bending moments in the narrow top flange and torsion in the section in order to maintain equilibrium. It is found that the construction condition controls the design of the beam in cases where the imposed load applied to the composite section is small ($<3\,kN/m^2$).

The out-of-balance moments on the beam are resisted by the connections at the ends of the beam. Four-bolt full-depth end plate connections usually achieve adequate torsional and shear resistance. When the beam is propped during construction, combined bending and torsional effects are much reduced, and do not influence the design of the beam.

3.8.4 Ultimate limit state

The steel beam is designed to act compositely with the concrete slab at the ultimate limit state, partly due to the patented raised rib pattern rolled into the top flange of the section. The beam resists the bending and shear due to the factored loads applied over the entire supported

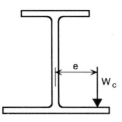

Load case 1 : Bending and torsion

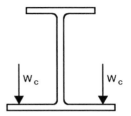

Load case 2 : Bending

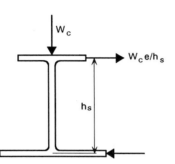

Forces on beam in load case 1

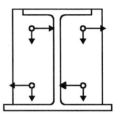

Forces at end connections in load case 1

Figure 3.30   Loads developed in the construction stage.

area. Out-of-balance loads do not cause additional stresses in the composite beam because of the restraint provided by the torsional and bending stiffness of the slab at the ultimate limit state.

As in conventional design to BS 5950: Parts 1 and 3, partial factors for loads of 1.6 and 1.4 are applied to the imposed and dead loads respectively to obtain the maximum factored bending moment and shear force acting at the ultimate limit state.

The bending resistance of the composite section is established on plastic analysis principles using a steel design strength of $p_y = 345 \, \mathrm{N/mm^2}$ (see design properties of steel sections in Section 3.8.3). The effective breadth of slab considered to act with each beam is taken as beam span/8 (or half of the value in conventional composite design to BS 5950: Part 3). This is done in order not to over-estimate the degree of composite action which is limited by the modest amount of reinforcement that is placed in the slab over the beams.

The compressive force that can be generated in the concrete is also limited by the longitudinal shear bond between the steel beam and the concrete encasement which is enhanced by the raised rib pattern on the top flange of the beam. The design value of this bond stress is $0.6 \, \mathrm{N/mm^2}$ which is considered to act over an effective perimeter equal to the top flange and web of the beam. Because shear bond failure is not fully ductile, the shear flow along the beam is considered to be linear rather than uniform, which would be the case for conventional shear connectors.

The actual degree of composite action and shear bond strength is demonstrated by full-scale load tests. It is found that a significant increase in the moment resistance of the steel section can be obtained as a result of composite action. However, it may not be necessary to utilise all of this increased resistance at the ultimate limit state because of the importance also of the serviceability and fire limit states.

The detailed design procedures for ASB sections are presented in Section 3.8.7.

3.8.5 *Serviceability limit state*

The serviceability limit state is largely concerned with the control of deflections under unfactored loads (i.e. working loads). Elastic section properties are used to calculate deflections, which comprise two components:

- the deflection of the steel beam under the self-weight of the slab,
- the deflection of the composite beam under imposed loads.

The total deflection is the sum of these two components. It is recommended that the total deflection of internal beams should be limited to a maximum value of span/200, so that deflections are not visible and do not affect the attachment of the ceiling elements or raised floor. When the total deflection exceeds this limit (or a maximum value of 50 mm) it may be necessary to consider propping or precambering of long-span beams.

The imposed load deflection may be more critical than the total deflection, especially for edge beams. Conventionally, the deflections due to imposed loads alone are limited to a maximum value of span/360 to avoid cracking of partitions, etc. It may be necessary to adopt more severe deflection limits for edge beams that support brittle cladding, such as brickwork. Typically, an imposed load deflection limit of span/500 and a total deflection limit of 25 mm are adopted for edge beams.

The stiffness of a steel ASB beam is established from its second moment of area. The stiffness of the composite ASB beam is established by taking account of the area of the concrete encasement and the effective breadth of concrete slab. The stiffness is based on the uncracked section property, in which the concrete area is divided by an appropriate modular ratio (ratio of the elastic moduli of steel to concrete). The use of the uncracked composite stiffness is justified by tests and by the adoption of a conservative value of effective slab breadth. The uncracked stiffness is only slightly higher than the cracked value, and its use greatly simplifies checks at the serviceability limit state.

No stress checks are required for ASB sections at the serviceability limit state because tests have shown that the beam behaviour is essentially elastic until well above working loads.

The same uncracked stiffness may be used to establish the natural frequency of the beam, as explained in Section 2.2.13. The natural frequency is calculated based on the quasi-static deflection of the composite beam subject to all permanent loads (taken as self-weight, other dead loads and 10% of imposed loads).

The lower design limit to the natural frequency of the beam is 4 cycles/second which is chosen so that resonant vibrations do not occur under the normal range of activities in office buildings. This limit may be reduced to 3 cycles/second in industrial buildings or car parks, but should be increased to 5 cycles/second in buildings (such as hospitals) that require further control of deflections and vibrations.

3.8.6 *Fire limit state*

The three current ASB sections have been developed to achieve 60 minutes fire resistance in slim floor construction without requiring additional protection of the exposed bottom flange. These sections have thicker webs so that they retain more of their bending resistance than conventional Slimflor beams in fire conditions. This is because the bottom flange may reach temperatures of over 800°C, whereas the web and upper flange and the concrete encasement are much cooler, and are fully effective.

All Slimflor and ASB beams can achieve 30 minutes fire resistance without further checks. The critical case is that for 60 minutes fire resistance where ASB sections do not require protection in the majority of applications (see below). For longer periods of fire resistance, additional protection to the bottom flange of the ASB section is required.

The moment resistance of the composite ASB section in fire conditions is established on plastic analysis principles, taking account of the strength reductions of the bottom flange, web and concrete encasement based on measured temperature distributions. These calculations have been carried out for all the ASB sections.

The 'load ratio' is defined as the load or moment applied at the fire limit state divided by the moment resistance at the ultimate limit state.

BS 5950: Part $8^{(2)}$ permits the use of a reduced partial factor for variable imposed loads. The maximum load ratio depends also on the degree of composite action that is utilised at the ultimate and fire limit states. A load ratio of 0.5 is generally appropriate for offices or buildings of variable occupancy, increasing to 0.6 for warehouses or industrial buildings.

When protection is required for a longer fire resistance period, the thickness of fire protection is established using standard tables for spray, board and intumescent coating products given in the ASFPCM/SCI publication.[11]

3.8.7 *Design procedures for ASB sections*

The following procedures are presented for calculation of the section properties and bending resistance of the ASB sections. General equations are presented in Appendix B for composite asymmetric sections. Section properties for the ASB section range are presented in Table 3.3, which also take account of the root radii between the flanges and webs. The method of analysis for the ASB section is similar to many aspects of Slimflor design already demonstrated in previous sections of this publication. Hence, only the key design parameters associated with ASB design will be given.

3.8.8 *Moment resistance of composite section*

The moment resistance of the composite ASB section takes into account the compression transferred to the concrete slab and is calculated using conventional plastic analysis principles according to BS 5950: Part 3, as shown in Fig. 3.31. The plastic neutral axis of the composite section generally falls into the lower part of the web so that all of the slab is in compression. Therefore, equilibrium is satisfied when the compressive resistance of the concrete slab and top flange is equal to the net tensile force in the steel section.

The degree of composite action that can be developed depends on shear bond action between the steel section and the concrete encasement, and also the amount of reinforcement placed transverse to the beam (see Section 3.8.12). The raised rib pattern on the top flange enhances the shear bond action considerably. The concrete contained between the flanges and end diaphragms is effectively held in place and achieves good composite action with the beam.

The effective breadth of the slab is limited to a maximum of beam span/8 which is half of the value adopted for a normal composite beam. This conservatism means that the benefit of composite action is

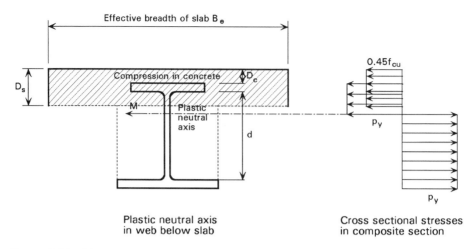

Plastic neutral axis
in web below slab

Cross sectional stresses
in composite section

Figure 3.31   Moment resistance of composite ASB section.

under-estimated and takes account of the modest amount of transverse reinforcement that is used in practice. Furthermore, the slab depth over the steel beams should be a minimum of 30 mm and a maximum of 60 mm in order that the proposed design is within the limits of the test information. If the slab depth over the beam is less than 30 mm, composite action due to the top flange should be ignored.

3.8.9 *Full shear connection*

In order to achieve full shear connection in a composite beam, it is sufficient to ensure that the compressive resistance of the cross-sectional area of the solid concrete above the steel decking and within the effective breadth can be developed. The compressive resistance of the slab is given in clause 4.4.2 in BS 5950: Part 3 as

$$R_c = 0.45 f_{cu} D_s B_e$$

where B_e is the effective breadth of slab $= L/8$
D_s is the solid slab depth above the decking
f_{cu} is the compressive cube strength of concrete.

The steel section is partially contained within the slab depth, D_s, and therefore the upper part of the section also acts in compression with the slab.

The plastic neutral axis of the composite section is established by equating tension and compression forces, based on rectangular stress blocks acting in the steel and concrete. The most common design case is where the plastic neutral axis lies in the steel web below the solid concrete slab, as shown in Fig. 3.31. Therefore, only the bottom flange and the lower part of the web resist tension. In this case, the depth of the steel web in compression is given by the formula

$$y_s = d\left(\frac{R_b + R_w - R_t - R_c}{2R_w}\right)$$

where R_t, R_b and R_w are the tensile resistances of the top flange, bottom flange and web of the ASB section respectively
    d is the depth of the web between the flanges.

The corresponding depth, y_c, of the plastic neutral axis from the *top* of the slab is given by

$$y_c = y_s + T_t + D_c$$

where D_c is the top cover to the steel section (30 mm $\leq D_c \leq$ 60 mm)
    T_t is the thickness of the top flange.

The plastic moment resistance of the composite section is established by taking moments of the various stress blocks around the plastic neutral axis position. For the case shown in Fig. 3.31, where the plastic neutral axis lies in the web below the solid concrete, the plastic moment resistance is given by the formula

$$M_c = R_t(y_s + 0.5T_t) + R_b(d - y_s + 0.5T_b)$$

$$+ R_c(y_s + T_t + D_c - 0.5D_s)$$

$$+ R_w\left\{\left(\frac{y_s}{d}\right)^2 - \left(\frac{y_s}{d}\right) + 0.5\right\}d$$

where all the parameters are defined as previously and T_b is the thickness of the bottom flange.

The plastic moment resistance for this, and other plastic neutral axis positions, are presented in Appendix B in terms of y_c rather than y_s. In theory, five design cases (corresponding to fire positions of the plastic neutral axis) exist, but in practice only two or three of them normally apply for composite ASB sections.

3.8.10 *Longitudinal force in the slab*

The longitudinal shear force transfer is assumed to occur by shear bond stresses acting uniformly around the upper flange and both sides of the web of the ASB section. For beams subject to uniformly distributed load, the maximum compressive force, F_{sb}, that can act in the slab at mid-span of the beam is obtained from consideration of elastic shear flow along the beam, as follows

$$F_{sb} = (B_t + T_t + d - 0.5t_w)\frac{L}{2}f_{sb}$$

where f_{sb} is the average shear bond strength around the top flange and web.

If there is no, or insufficient, cover to the top flange, then the term in B_t is set to zero, implying that no longitudinal shear force is transferred by the top flange. The force F_{sb} is also dependent on the amount of transverse reinforcement (see Section 3.8.12).

This shear bond action is illustrated in Fig. 3.32. The force f_{sb} has a design value of $0.6\,\text{N/mm}^2$, as justified by full-scale tests for ASB sections with their raised pattern rolled into the top flange. For other

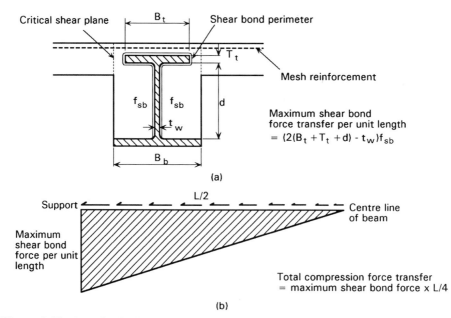

(a)

(b)

Figure 3.32   Longitudinal shear connection due to shear bond action in a uniformly loaded beam. (a) Shear bond transfer around the internal surface of the section. (b) Elastic shear transfer along the beam.

load cases, composite action is ignored unless the elastic shear flow is less severe than the uniformly distributed load case (e.g. a beam with a central point load is acceptable).

For the case where the plastic neutral axis lies below the solid concrete slab, full *shear connection* is achieved when $F_{sb} \geq R_c$. Therefore, the full compressive resistance of the concrete slab is developed, and the plastic moment resistance is obtained from the above equation for M_c. When the plastic neutral axis lies within the solid concrete, the mathematical definition of full shear connection is not so straightforward, but the equation for F_{sb} is always conservative.

For cases of full shear connection, the moment resistance of a composite ASB section is typically 1.3–1.5 times that of the ASB section which greatly improves its resistance to imposed loads. However, other criteria often control the design of these beams, and the bending resistance is not generally the limiting factor.

3.8.11   Partial shear connection

Partial shear connection exists when there is insufficient longitudinal shear bond to develop the plastic bending resistance of the composite section. For the case where the plastic neutral axis lies below the solid concrete, this occurs when $F_{sb} < R_c$.

In this case, the plastic moment resistance of the composite section is reduced, and the calculation procedure is modified by taking a reduced compression force in the concrete slab, given by F_{sb}. Therefore, the term in R_c may be replaced by F_{sb} in the calculation procedure for y_s and M_c, as given in Section 3.8.9.

Alternatively, a simple linear interaction formula may be used to determine the design moment resistance, M_d, of the composite section for partial shear connection, as follows

$$M_d = M_s + \frac{F_{sb}}{R_c}(M_c - M_s)$$

where M_s is the moment resistance of the steel section ($= S_x p_y$)
M_c is the moment resistance of the composite section for full shear connection
F_{sb}/R_c is the degree of shear connection.

In all cases, the design moment resistance should be greater than or equal to the applied moment at the ultimate limit state.

The typical relationship between the moment resistance of a composite ASB section and the degree of shear connection is illustrated in

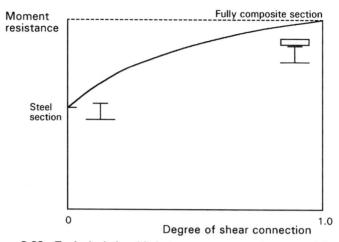

Figure 3.33 Typical relationship between moment resistance and degree of shear connection for ASB and composite ASB beams.

Fig. 3.33. The linear interaction formula for M_d is conservative with respect to the true interaction.

According to clause 5.5.2 of **BS 5950: Part 3**, the minimum degree of shear connection, defined by the parameter F_{sb}/R_c, should always be greater than 0.4. This limit is generally satisfied in practice by ensuring that the maximum depth of concrete over the beams is 60 mm (in order to limit the value of R_c).

Additional bars that are attached to, or passed through, the steel section also enhance the degree of shear connection. This may be the case for beams where the concrete is placed level with the top flange, or for edge beams. The shear strength of these bars is based on $0.7 f_{ys}$ (see Section 3.8.12). The ductility of the bar reinforcement is such that the combined shear resistance of all the bars between the points of zero and maximum bending moment may be developed.

The full-scale tests carried out at City University have demonstrated that the shear connection due to shear bond action is robust, and the moment resistance of the composite ASB section may be calculated using conventional plastic theory. However, it may be necessary to reduce the longitudinal shear force if there is insufficient transverse reinforcement (see Section 3.8.12).

3.8.12   *Transverse reinforcement*

Mesh reinforcement is placed over the top of the beam and transfers longitudinal shear forces due to composite action into the slab.

The longitudinal shear resistance of the slab per unit length is given in clause 5.6.3 of BS 5950: Part 3 as

$$v_r = 0.7 A_{sv} f_{ys} + 0.03 \eta A_{cv} f_{cu}$$

where: A_{sv} is the cross-sectional area of the transverse reinforcement per unit length of the beam
f_{ys} is the yield strength of the reinforcement
A_{cv} is the area of one shear plane per unit length of the beam which includes the area of both the concrete topping and rib

$$\eta = \begin{cases} 1.0 \text{ for normal weight concrete} \\ 0.8 \text{ for lightweight concrete.} \end{cases}$$

Sufficient transverse reinforcement is placed when the longitudinal shear resistance across one shear plane (shown in Fig. 3.32) satisfies the following formula

$$v_r \geq 2 \frac{F_{sb}}{L} \frac{(B_e - B_b)}{B_e} r_s$$

and

$$F_{sb} \leq R_c$$

where r_s is the reduction factor due to average shear across one rib
B_b is the width of bottom flange (solid area of concrete)
B_e is the effective breadth of slab $(= L/8)$
L is the beam span.

The reduction factor, r_s, takes into account the average longitudinal shear force across one rib at the end of the beam and may be taken as $(L - 2b_s)/L$, where b_s is the rib spacing $(= 600\,\text{mm})$. Conservatively, r_s is 0.85.

This formula assumes that the local transfer of longitudinal shear force is established from elastic shear flow along the beam. If there is insufficient transverse reinforcement, F_{sb} may be reduced until the above equation for v_r is satisfied. This value of F_{sb} should be used to calculate the moment resistance of the composite section, according to the equations for F_{sb} or M_d. The fire design is not significantly influenced by the precise value of F_{sb}, as full shear connection is achieved in fire conditions.

The minimum depth of concrete over the beam is 30 mm so that the mesh reinforcement can be placed with sufficient top cover. In cases where the beam is subject to torsion, additional bars should be anchored or attached to the beam, or located through pre-punched holes in the beam web. These bars may also act as shear connectors and as transverse reinforcement to provide for longitudinal shear transfer.

The above equation for v_r is conservative as it assumes that the worst condition occurs at the ends of the beam and ignores the effect of controlled slip due to yielding of the transverse reinforcement. It also ignores the contribution of the bar reinforcement in the ribs and the shear bond action of the end diaphragms. The full-scale tests carried out at City University showed that A142 mesh reinforcement over the beam is sufficient to prevent non-ductile failure. It is argued that this design check is not necessary provided that the concrete cover over the beam is between 30 mm and 60 mm and at least A142 reinforcement is used.

3.8.13 *Elastic properties*

The relevant elastic property for stiffness calculations is the second moment of area (inertia) of the composite ASB section. Tests have shown that the uncracked inertia of the composite ASB sections may be used, provided that the stiffness of the concrete encasement is reduced by dividing its area by a suitable value of modular ratio, α_e. The modular ratio is the ratio of the elastic modulus of steel to concrete, and may be taken as 10 for normal weight concrete and 15 for lightweight concrete for beams subject to predominantly variable load, such as in offices. A higher value of modular ratio should be used for applications in which the loading is semi-permanent.

The composite cross-section used for calculation of elastic section properties is illustrated in Fig. 3.34. It includes both the concrete slab

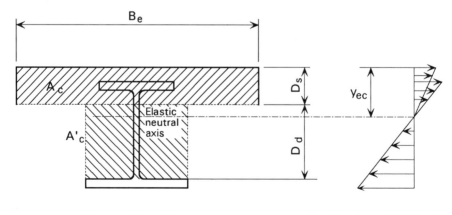

Composite section   Elastic stress distribution

Figure 3.34   Elastic section property of composite ASB section.

area, A_c, and concrete encasement, A'_c. The elastic neutral axis of the composite section is first calculated using conventional elastic principles. The second moment of area is then established by taking moments about the elastic neutral axis. The elastic properties of the ASB sections are presented in Table 3.3.

It is often found that the stiffness of the composite ASB section is 1.5–2 times that of the ASB section, which greatly reduces its deflections under imposed loads. However, in unpropped construction, the total deflection under all loads should also take into account the deflection of the steel beam subject to the self-weight of the floor slab. A total deflection limit of span/200 is appropriate for internal beams, and this limit often controls the design of ASB beams with thicker slabs.

3.9 RHS Slimflor edge beams

3.9.1 *Edge beams in Slimflor construction*

The first consideration for edge beam design is to determine the form of construction that is appropriate architecturally. The selected form of construction for the edge beam will dictate the design concept. For example, if a downstand beam is acceptable with regard to the cladding details, a more traditional design approach can be adopted. The decking can rest on the top flange of the beam and the vertical loads may be assumed to act through the shear centre of the beam.

In the past, this option has proved to be a popular method of construction because it has the advantage of eliminating the torsional effects on an open beam section. However, the use of downstand beams has the disadvantage of increasing the construction depth, which in turn can affect the architectural details and appearance of the building.

However, for edge beams, the use of a conventional Slimflor beam with an open section does give rise to some problems. The section is torsionally weak and flexible. The eccentric loads from the slab can lead to the use of relatively heavy sections to avoid excessive twist. To provide a more efficient solution for edge beams in this form of construction, the rectangular hollow section (RHS) Slimflor beam has been developed, as shown in Fig. 3.35. The good torsional properties of the RHS section ensure that stresses and movements due to eccentric loads are minimised.

The RHS Slimflor beam can readily be used with RHS perimeter columns, leading to more slender walls and, where the steel is exposed, to better appearance.

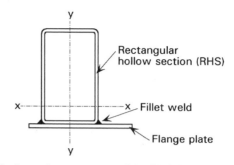

Figure 3.35 Basic steel components of the RHS Slimflor beam.

The additional benefits of RHS Slimflor construction are:

- Good torsional properties of the edge beams.
- Smooth external face providing easier connections for brackets, cladding, etc.
- Efficient member size, leading to reduced steel weights.
- Good fire resistance.
- Wide range of RHS section sizes and thicknesses.

With the introduction of RHS edge beams in Slimflor construction, there is an opportunity to simplify cladding details and to offer an architecturally improved design in which the edge beam can be exposed. The provision for subsequent movement between the frame and cladding may be reduced.

3.9.2 *Form of construction*

The RHS Slimflor edge beam

The RHS Slimflor edge beam uses a steel plate welded to the underside of a standard RHS section. The bottom plate projects on one side, where it supports the floor slab, but has only a small projection on the other side to facilitate fillet welding and the attachment of cladding (see Fig. 3.35).

The RHS Slimflor beam is particularly well adapted to edge beam applications because of its good torsional properties and smooth external appearance.

3.9.3 *RHS Slimflor edge beams acting non-compositely with floor slab*

Figure 3.36 shows typical details for a RHS Slimflor edge beam acting non-compositely. In principle, the beams are designed to be unpropped

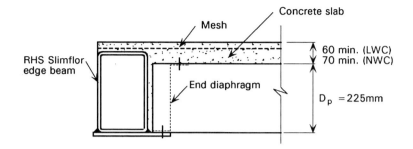

Cross-section through non-composite Slimflor RHS beam - Type A

Figure 3.36   Non-composite edge beam.

during construction, although there may be circumstances where propping is used to provide the minimum steel beam size for a given
span. Grade S355 steel (formerly grade 50) and lightweight concrete
are the preferred materials. However, where serviceability criteria
control the design, use of grade S275 steel (formerly grade 43) is more
cost effective.

3.9.4   *RHS Slimflor edge beams acting compositely with floor slab*

Figure 3.37 shows typical details for a RHS Slimflor edge beam acting
compositely with the floor slab. Again, the beams are usually designed
to be unpropped during construction; grade S355 steel and lightweight
concrete are the preferred materials. Although the bottom plate of the
edge beam does not usually project much beyond the web of the RHS,

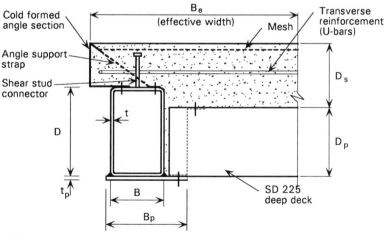

Cross-section through composite Slimflor RHS beam - Type B

Figure 3.37   Composite edge beam.

in composite applications it is often necessary for the slab to project over the RHS in order to provide the required edge distance for the stud shear connectors (i.e. $6 \times$ stud diameter). Transverse reinforcement in the form of U bars is required in the slab in this case.

3.9.5 *Construction details*

Beam-to-column connections

General

RHS edge beams are well suited to use with RHS columns. In the details presented below the perimeter columns are generally shown as RHS sections, although connections can also be made to UC columns.

Non-composite edge beams

Figure 3.38 shows typical structural details for non-composite RHS edge beams.

Figure 3.38(a) shows the case where the deck is orientated perpendicular to an RHS edge beam. The edge beam has a 15 mm thick plate welded to the underside that projects 100 mm from the face of the RHS. This 100 mm projection acts as a suitable support to the decking. If the bottom flange plate projects 100 mm from one side and, say, 10 m the other side, then the beam can be welded (flange plate to RHS) without the need for turning the section. However, this slight projection may affect the cladding details and therefore an alternative detail requiring welding to the underside may be used.

A full depth end plate that projects above and below the beam is welded to the section. Conventional bolts may be used to connect the end plate to a UC column. This type of connection is effective in resisting torsional moments from the beam.

Flowdrill or Hollo-Bolt type bolting techniques are used to connect the edge beam to a RHS column. Adopting this form of fabrication provides a very neat solution for the beam to RHS column connection. However, due consideration must be given to the erection tolerances, i.e. practical buildability of the system.

The tie member shown in Fig. 3.38(a) is confined within the slab depth and provides a support for the decking. This tie member can take the form of any suitable structural section provided it can withstand part of the floor load in the construction stage, and will not deflect significantly between the columns. This arrangement would be suitable for buildings greater than four storeys and must satisfy code requirements for robustness. For reasons of cost and practicality, the tie member is likely to be fabricated from a structural T section.

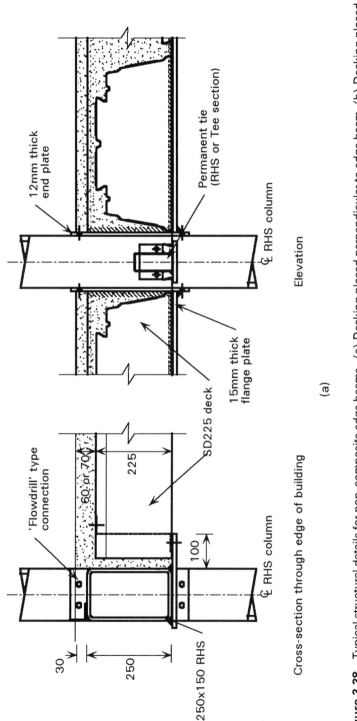

12mm thick
end plate

Permanent tie
(RHS or Tee section)

15mm thick
flange plate

℄ RHS column

Elevation

'Flowdrill' type
connection

SD225 deck

60 or 70

225

100

30

250

250x150 RHS

℄ RHS column

Cross-section through edge of building

(a)

Figure 3.38 Typical structural details for non-composite edge beams. (a) Decking placed perpendicular to edge beam. (b) Decking placed parallel to edge beam.

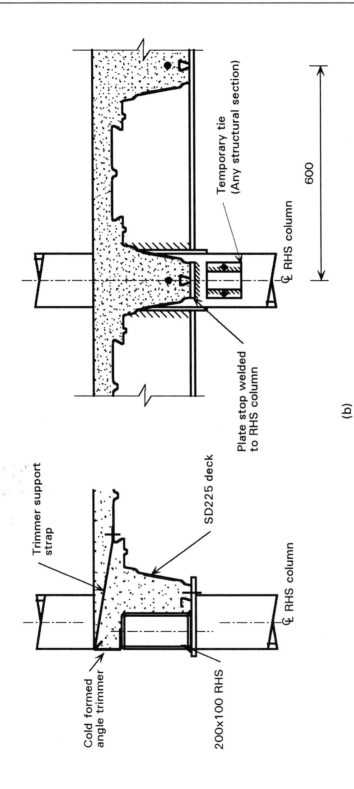

Figure 3.38 (*continued*).

Figure 3.38(b) shows the case where the decking runs parallel with the edge beam. This figure also shows an alternative tie detail for buildings up to four storeys high. The underside of the deck trough is supported by a separate plate welded to the column. The main advantage of using this method of construction is that the decking can be placed continuously at the column position and is not affected by the tie member.

Before the in situ concrete can be placed (normally by pumping), cold formed angle edge trimmers have to be connected to the RHS edge members. These edge trimmers require support straps at approximately 400 mm centres. Also, bar reinforcement (for fire resistance purposes) is placed in each trough and a layer of mesh (A142) is positioned towards the surface of the slab with the minimum amount of cover.

Figure 3.39 shows an isometric projection of the details in Fig. 3.38(a) in order to further illustrate the method of construction.

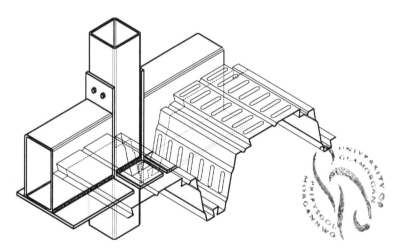

Figure 3.39   Isometric projection of RHS edge beam and deep decking.

Composite edge beams

Figure 3.40 shows typical structural details for a composite RHS edge beam, where the decking is placed perpendicular or parallel to the edge beam. In this case, the basic construction details are similar to those shown in Fig. 3.38, the main difference being the use of 19 mm diameter × 70 mm long shear connectors (studs) welded to the upper surface of the RHS edge beam. In this case, the RHS wall thickness

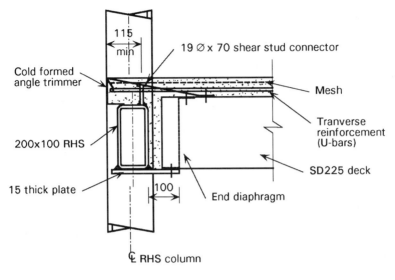

(a)

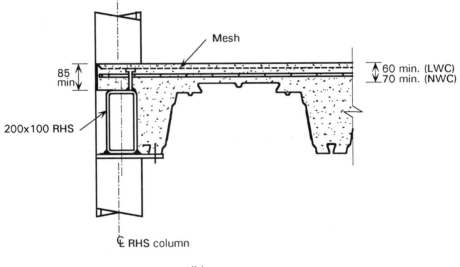

(b)

Figure 3.40   Typical structural details for composite edge beams. (a) Decking perpendicular to edge beam. (b) Decking parallel to edge beam.

must be at least 8 mm, which is necessary to prevent burn through of the RHS. In addition to the mesh reinforcement, transverse reinforcing bars (U bars) are required to ensure a smooth transfer of forces from the studs to the slab.

3.9.6 *Support of cladding*

Cladding panels transmit lateral forces, arising from wind loading and other effects, to the building structure. Where cladding panels are connected to edge beams of open section, problems of transverse bending and torsion of the beam can arise. These are particularly severe when the top edge of a storey-height cladding panel is attached to the lower flange of the beam. Further information dealing with curtain wall connections can be obtained from the SCI publication *Curtain wall connections to steel frames.*[12]

Edge details in buildings often mean that it is difficult to avoid making connections to the lower flange of these open edge beams. Also, secondary steelwork may be needed to provide lateral support to open sections; this can take various forms such as wind restraint beams or diagonal struts connecting the lower flange of the beams to the soffit of the floor slab. Both of these options have significant disadvantages in terms of cost or installation time.

For RHS edge beams it is unnecessary to stiffen the beams locally because, due to their high torsional stiffness and strength, they are capable of resisting such actions. The RHS edge beam provides a structural solution that is simple, cost effective and architecturally pleasing.

Strongbacks and integral panels

Strongbacks
Stone veneer cladding and other thin materials, or materials with inadequate spanning capabilities, may be supported on frames known as strongbacks. Typically, these are aluminium, steel or stainless steel sub-frames onto which panels are screwed or bolted. One strongback may receive several cladding panels, enabling cladding such as stone to be used in relatively small, manageable sizes appropriate to the structural properties of the veneer. Since the strongback resists the structural loads, it is usually possible to design panels to span from column to column.

Integral panels
Heavy reinforced concrete cladding panels, known as integral panels, are fixed to the primary structure in a similar way to strongback panels. Integral panels may be either top hung or bottom supported and typically bear onto the floor slab using either a boot (a projecting nib or ledge) or bolted on brackets (normally based upon an angle or series of angles).

Integral panels may be clad in other materials such as stone. The increased weight of these panels can lead to greater lateral loading of the edge beams, particularly where there is eccentric loading.

Supports

Support brackets for strongbacks or integral panels may be attached to columns, to the floor edge, or to a combination of both. Panel head brackets may be fixed directly to channels or to other fixing points welded to the lower flange of the edge beam.

Figure 3.41 shows a support arrangement for a typical strongback cladding panel. Top of slab brackets are cast into the floor, having been positioned using 'chair' type brackets. The top of the fixing channels (or equivalent devices) is flush with the top of the slab. Figure 3.42 shows the same detail in isometric and plan views.

Stick system

Stick systems used in cladding comprise linear vertical and horizontal members. Vertical members (mullions) are nominally continuous, while horizontal members (transoms) are normally discontinuous. Conventionally, they are built in situ by the cladding contractor rather than the steelwork fabricator.

Stick systems are generally used for lighter cladding materials such as steel, aluminium and glass. The most common form of stick system

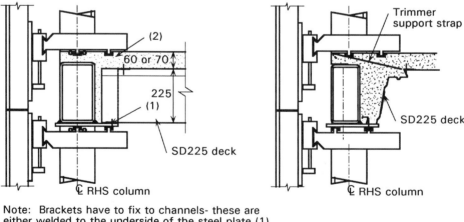

Note: Brackets have to fix to channels- these are either welded to the underside of the steel plate (1) or cast flush with the concrete (2).

(a)   (b)

Figure 3.41 Alternative cladding attachment details. (a) Decking perpendicular to angle beam. (b) Decking parallel to edge beam.

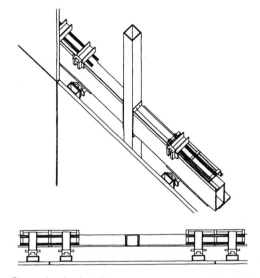

Figure 3.42 Strongback cladding system supported by RHS edge beams (isometric and plan views).

comprises rectangular panels restrained along four edges by extruded aluminium transoms and mullions.

Supports

Mullions in stick systems may be fixed either to columns or to edge beams. The fixing arrangement will generally allow vertical movement due to deflection of the floor, but must transmit lateral loads to the building structure.

RHS edge beams will have better resistance to torsional bending than will conventional downstand edge beams. Brackets should be fixed directly to either the columns or the edge beams using channels welded to the outside face of the steelwork, or other devices such as Flowdrill type connections.

Figure 3.43 shows the cross-section of an RHS attachment to a mullion and Fig. 3.44 shows the same detail in isometric and plan views.

Brickwork

Brickwork on multi-storey buildings has to be divided over its vertical height into a series of structurally independent bands in order to reconcile the differential movement that can occur between the structure and the cladding. Brickwork expands and contracts as a result of changes in temperature, experiences permanent moisture expansion and is susceptible to movement-related cracking.

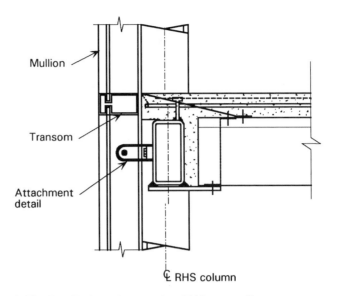

Mullion

Transom

Attachment
detail

℄ RHS column

Figure 3.43   Detail of attachment of an RHS to a mullion.

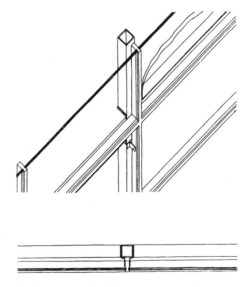

Figure 3.44   Stick system cladding supported by composite RHS edge beams (isometric and plan views).

Supports

In conventionally framed steel buildings, brickwork is supported off shelf angles. These stainless steel angles may be attached to the floor slab or to perimeter steelwork. Companies, such as Halfen, produce a comprehensive range of fittings for this purpose.

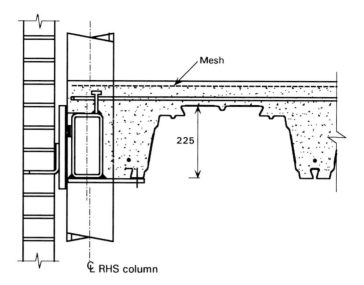

Figure 3.45   Detail of attachment of a RHS to brickwork.

Fixing to the floor edge can be difficult to achieve when the slab is thin or where the brickwork is heavy (particularly where there is a large eccentricity in the shelf angle loading). Fixing to steelwork can be problematic in that the steel beam can deflect as the brickwork is erected, compromising structural integrity.

Shelf angles may be fixed to RHS members either by bolting them to fixing channels welded to the face of the member, or using other devices such as Flowdrill type connections. RHS members are likely to have less tendency to deflect as a result of heavy eccentric loading than are conventional steel edge beams.

Figure 3.45 shows the cross-section of an RHS attachment to a brickwork support angle and Fig. 3.46 shows the same detail in isometric and plan views.

3.9.7 *Design of RHS Slimflor edge beams*

Basis of design

RHS Slimflor beams used with deep decking may be designed as either non-composite (see Fig. 3.38) or composite (see Fig. 3.40). Clearly, the different forms of construction give rise to different design considerations.

Both of these forms of construction may be used in cases where the decking is orientated perpendicular or parallel to the beam. The orientation of the decking to the edge beam has an influence on how the loads are transmitted to the edge beam.

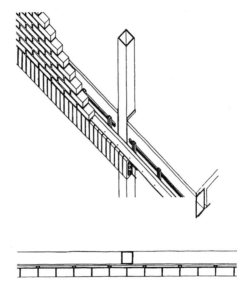

Figure 3.46 Brickwork cladding suppoted by composite RHS edge beam (isometric and plan views).

The general basis of design is similar to the Slimflor beam and the asymmetric Slimflor beam. The main design differences between these forms of construction are:

(1) Deflections of edge beams that are normally limited to span/500 under imposed loads, and span/250 under total load. A further deflection limitation for cladding and imposed loads of span/360 should also be considered. More strict deflection limits may be required for some forms of cladding.

(2) In the construction stage, the horizontal movement of the upper surface of the RHS is limited to span/500 in order to avoid possible problems with the cladding attachments. In addition, it is a requirement of BS 5950: Part 1 that due allowance should be made where deflections under serviceability loads could impair the strength or efficiency of the structure or its components, or cause damage to the finishings.

(3) The RHS Slimflor edge beam is considered to be laterally unrestrained at the construction stage, but restrained for the normal design.

3.9.8 *Torsional effects on the RHS edge beam*

As stated previously, the edge beam is subjected to loads that can cause twisting of the beam. The design assumptions for the two methods of construction are presented in the following sections:

Type A: non-composite beam

Deck orientated perpendicular to the edge beam

In the construction stage and in normal design, all floor loads (concrete, deck and construction load) are transmitted to the beam via the bottom flange plate. Cladding loads are assumed to be applied after the floor has been cast. The cladding loads partially counteract the torsion from the floor loads.

Deck orientated parallel to the edge beam

In the construction stage and in normal design, it is assumed that no loads are transmitted to the bottom flange plate. It could be argued that there will be a nominal load transmitted to the beam, but this is considered to be small and can be ignored. The eccentric cladding loads are resisted by torsion and bending in the RHS edge beam.

Type B: composite beam

Deck orientated perpendicular to the edge beam

As for the non-composite beam, the construction loads (the weight of the concrete, decking and construction load) are transmitted to the beam via the bottom flange plate. There is no cladding load at this stage.

However, for the normal design, it is assumed that the imposed floor loading and cladding loads act concentrically on the edge beams, i.e. no torsional effects are considered. This is because the transverse reinforcement is looped around the shear connectors welded to the section and out of balance moments are transferred directly as bending in the slab.

Deck orientated parallel to the edge beam

As for the non-composite beam, it is assumed that no construction loads are applied to the edge beam.

For the normal design, the cladding load is assumed to act concentrically to the edge beam and no imposed floor load is carried by the edge beam.

3.9.9 Design of non-composite beams (Type A)

Section classification

The cross-section will normally be classified plastic or compact to BS 5950 (Class 1 or 2 to EC3[10]). The relevant dimensions are shown in Fig. 3.47.

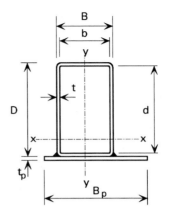

Figure 3.47   Section dimensions of composite beam.

The following width-to-thickness ratio limits are appropriate to a compact section.

Internal element of the compression flange $\dfrac{b}{t} \le 32\epsilon$

Web, generally $\dfrac{d}{t} \le \dfrac{98\epsilon}{\alpha}$

where $\alpha = \dfrac{2y_c}{d}$

$$\epsilon = \begin{cases} 1.0 & \text{for grade S275 steel} \\ 0.88 & \text{for grade S355 steel} \end{cases}$$

In most cases, the plastic neutral axis (pna) will be close to the bottom flange plate which means that the section should be assumed as having compression throughout the web. From the above expression it can be shown that, as $\alpha \approx 2$, then $y_c \approx d$. Hence, the limiting classification check becomes

$$\frac{d}{t} \le 39\epsilon$$

(this expression also applies to plastic and semi-compact sections).

This expression is likely to govern the section classification of rectangular sections. However, for square sections, the internal element of the compression flange ($b/t \le 32\epsilon$) will, in the majority of cases, govern for a compact section.

Lateral torsional buckling effects

Lateral torsional buckling (LTB) of a closed section is generally not considered due to its enhanced torsional properties. Nevertheless, there is a check for LTB in Appendix B of BS 5950: Part 1, Table 38. The basis of the LTB check is shown in Table 3.4.

Where the actual minor axis slenderness, L_E/r_y, is less than λ (as shown in Table 3.4), no further check for LTB is required and $M_b = M_c = S_x p_y$.

Biaxial stress effects in the flange plate

The design techniques for biaxial stresses in the flange plate are similar to the Slimflor beam. This action is illustrated in Fig. 3.48. (For further information see Section 3.6.6.)

Out-of-balance floor loading

Out-of-balance loading is illustrated in Fig. 3.49.

For the non-composite beam, it is assumed that all of the floor loads are eccentric to the edge beam. Where the decking is orientated perpendicular to the edge beam, the loads, W, are applied directly to the flange plate which, in turn, applies torsion to the beam. A rigorous method of analysis is used to combine the longitudinal bending effects with the torsion. This method of analysis is presented in the SCI publication *Design of members subject to combined bending and torsion*.[9] It is not appropriate to fully illustrate its use here, but the fundamental equations are explained.

Table 3.4   Limiting slenderness (λ) for hollow section of uniform wall thickness (BS 5950: Part 1, Table 38).

D/B[a]	λ
1	no limit
2	$\dfrac{350 \times 275}{p_y}$
3	$\dfrac{225 \times 275}{p_y}$
4	$\dfrac{350 \times 275}{p_y}$

[a]D and B are overall depth and width of section respectively.

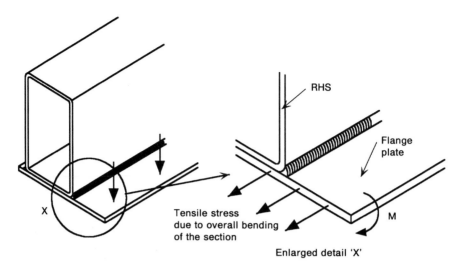

Figure 3.48   How loads are applied to the bottom flange plate.

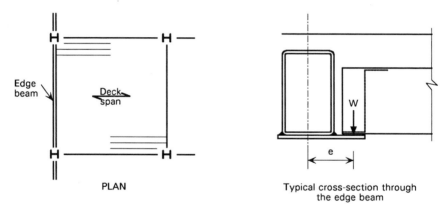

Figure 3.49   Typical plan layout showing edge beam arrangement.

There are two basic equations to be satisfied, as follows:

(1)   Buckling check

$$\frac{M_x}{M_b} + \left(\frac{\sigma_{byT} + \sigma_w}{p_y}\right)\left(1 + 0.5\frac{M_x}{M_b}\right) \leq 1.0$$

where M_x is the applied moment in major axis of the beam
M_b is the lateral torsional buckling resistance moment, but generally will be equal to M_c, plastic moment resistance.

$$\sigma_{byT} = \frac{M_{yT}}{Z_y}$$

$$M_{yT} = M_x \phi_{ULS}$$

ϕ_{ULS} is the angle of twist

σ_w is the warping stress (assumed to be zero for RHS).

(2) Capacity check

$$\sigma_{bx} + \sigma_{byT} + \sigma_w \leq p_y$$

where σ_{bx} is the longitudinal bending stress in the upperflange $= \dfrac{M_x}{Z_x}$.

The other terms are explained above for the buckling check.

Moment resistance of beam

The concrete that surrounds the beam is used for stiffness purposes only. Enhancement of the moment resistance of the beam by including the surrounding concrete with the steel member is difficult to justify unless additional shear connection is provided. It is for these reasons that the composite action is neglected at the ultimate limit state. The method of deriving the equation for the moment resistance of the steel member only using plastic analysis for the design of the cross-section is described below.

Figure 3.50 shows the position of the plastic neutral axis (pna) in the web, at a distance y_p below the centre line of the RHS. To analyse

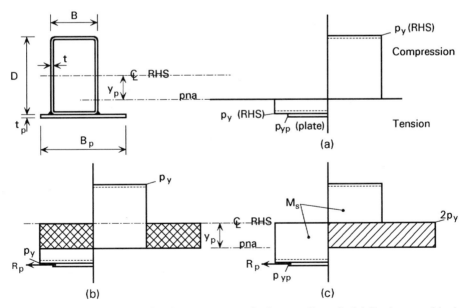

Figure 3.50 Stress blocks for the non-composite beam – Type A. (a) Basic stress blocks. (b) Addition of self-equilibrating stress blocks (shown by cross-hatching). (c) Modified plastic analysis.

the moment resistance from this diagram would involve some tedious calculations. In order to simplify the calculation, Fig. 3.50 also illustrates a standard method of rearranging the rectangular stress blocks.

The value of y_p can be found by equating the tensile resistance of the plate, R_p, to the increased compressive zone in the web. Hence

$$y_p = \frac{R_p}{4p_y t}$$

Moments are now taken about the mid-height of the RHS to determine the moment resistance, M_c, as follows

$$M_c = M_s + R_p \left(\frac{D}{2} + \frac{t_p}{2} \right) - \frac{R_p}{2} \left(\frac{R_p}{4t p_y} \right)$$

$$M_c = M_s + \frac{R_p}{2} \left[(D + t_p) - \frac{R_p}{4t p_y} \right]$$

where: $M_s = S_x p_y$ (moment resistance of RHS section)
$R_p = B_p t_p p_{yp}$
B_p is the width of the plate
t_p is the thickness of the late
p_{yp} is the design strength of the plates.

For other positions of the pna, M_c can be calculated using the formulae given in Appendix B.

The basic stress blocks are shown in Fig. 3.50(a). Addition of self-equilibrating stress blocks (marked with cross-hatching) in Fig. 3.50(b) leads to the modified plastic analysis in Fig. 3.50(c).

Vertical shear capacity

The vertical shear capacity, P_v, of the steel member is determined using BS 5950: Part 1 as follows:

$$P_v = 0.6 p_y A_v$$

where A_v is the shear area taken as $\left(\dfrac{D}{D+B} \right) A$.

Vertical shear can influence the moment resistance of the beam. This occurs where high shear and moment co-exist at the same position within the span. Simply supported beams with one or two point loads are good examples of cases where this occurs. Point loads are not covered by this book as its scope is restricted to simply supported edge beams with uniform loading only. For this condition, any influence the shear might have on the moment resistance is considered as minimal.

3.9.10 *Design of composite beams (Type B)*

The main assumptions and construction stage design are as for the non-composite beam (see Section 3.9.9), except for the torsion design in the normal stage. It is assumed that the floor loads are transferred to the composite section and not via the bottom flange plate, hence eliminating the torsion (out-of-balance loading) in the normal design stage. Other design parameters are defined below.

Effective width of compression flange

The effective width of slabs, B_e, will vary at the ultimate and serviceability limit states but a design value of $(\frac{L}{8} + \frac{B}{2})$ for edge beams has been adopted for both cases in BS 5950: Part 3: Section 3.1. The value of B_e should not exceed the distance between beam centres, although this is not likely to occur in practice.

Moment resistance of beam

The moment resistance of a composite RHS Slimflor edge beam is dependent on the degree of shear connection acting between the concrete and steel beam.

Full shear connection
Full shear connection occurs where the number of shear connectors provided is at least equal to that required to develop the full resistance of the concrete or the steel member. This will generate the maximum moment capacity of the cross-section. Full shear connection exists when:

$$R_q \geq R_c \quad \text{and} \quad R_q \geq R_s + R_p$$

where R_q is given by the number of shear connectors in the half span (between the positions of zero and maximum moments) multiplied by the resistance of each shear connector, determined from Table 5 of BS 5950: Part 3, reduced by a factor of 0.8 (and further by 0.9 for lightweight concrete).

Note that when $y_c = D_s$, then $R_q = R_c$.

The moment resistance for partial shear connection can be determined from other positions of plastic neutral axis, and formulae are given in Appendix B. In using these formulae

$$R_c = B_e(D_p + D_s - D)\,0.45 f_{cu}$$

$$R_s = A p_y$$

where A is the cross-sectional area of the RHS section.

Partial shear connection

Partial shear connection design may be used when there is an excess of moment resistance compared to the applied factored moment. In this situation, BS 5950: Part 3: Section 3.1 allows a reduction in the number of shear connectors that are needed (to a maximum of 40% for spans up to 10 m). For spans between 10 m and 16 m, the following relationship should be satisfied

$$\frac{N_a}{N_p} \geq \frac{L-6}{10} \quad \text{but} \quad \frac{N_a}{N_p} \geq 0.4$$

where N_a/N_p is the degree of shear connection
       L is the beam span (in metres).

The principle is that the number of shear connectors may be reduced so that the force transferred, R_q (see above), is sufficient to provide the required moment resistance. In this case, the force in the slab is R_q (not R_c) and the moment resistance may be determined from the stress blocks in Fig. 3.51.

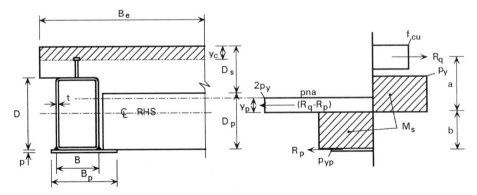

Figure 3.51 Typical cross-section through composite slab with partial shear connection.

Position of plastic neutral axis (above mid-height of RHS) is

$$2(2p_y t y_p) = R_q - R_p$$

$$y_p = \frac{R_q - R_p}{4p_y t}$$

Taking moments about the mid-height position of the RHS to find the moment resistance, M_c

$$M_c = M_s + R_q a + R_p b + (R_q - R_p)\frac{y_p}{2}$$

$$\therefore \quad M_c = M_s + R_q a + R_p b - \frac{(R_q - R_p)^2}{8p_y t}$$

where $M_s = S_x p_y$

$$a = D_s + D_p - \frac{D}{2} - \frac{y_c}{2}$$

$$y_c = \frac{R_q}{0.45 f_{cu} B_e}$$

$$b = 0.5(D + t_p).$$

Deflections

To provide general guidance on edge beam deflections is difficult because limiting deflection will largely depend on the methods adopted for the edge beam construction as for the cladding design. The following recommendations are intended as a guide and apply to buildings of general usage using curtain walling or masonry cladding. In this case, the edge beam deflection is limited to span/500 under imposed loads and span/250 under total loads. A further deflection limitation for cladding and imposed loads of span/360 should be considered.

In the construction stage, the horizontal movement of the top of the RHS is limited to span/500. In addition, it is a requirement of BS 5950: Part 1 that due allowance should be made where deflections under serviceability loads could impair the strength or efficiency of the structure or its components or cause damage to the finishings. It may be necessary to adopt more severe limits for some forms of cladding.

Transverse reinforcement

Transverse reinforcement is used to ensure a smooth transfer of the longitudinal force at the ultimate limit state (via the shear connectors) into the slab without splitting the concrete. The potential shear failure plane through the slab lies on only one side of the shear connectors, as shown in Fig. 3.52.

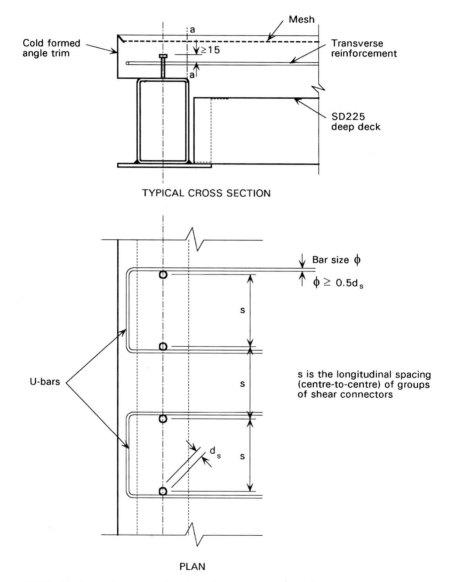

Figure 3.52 Typical edge beam details and assumed shear failure planes at the ultimate limit state.

The requirements of this section apply only to edge (L) beams where the slab edge is less than 300 mm from the nearest row of shear connectors. In order to develop their full resistance, the shear connectors must be less than $6d_s$ from the slab edge. Additional transverse reinforcement should be placed in the form of U-bars located below the top of the shear connectors. The diameter of the U-bars should not be less than $0.5d_s$. Detailing rules are presented in Fig. 3.52. There may be cases where it is not possible to observe this $6d_s$ limit. No code guidance is given on this case, but it is conservative to reduce the design resistance of the shear connectors by 50%, provided the edge distance is at least $3d_s$ and U-bars are used.

The shear resistance per unit length of each plane along the beam, v_r, is given by

$$v_r = 0.7 A_{sv} f_y + 0.03 \eta A_{cv} f_{cu}$$

but $v_r \leq 0.8 \eta A_{cv} \sqrt{f_{cu}}$

where f_{cu} is the characteristic cube strength of the concrete in N/mm², but not greater than 40 N/mm²

$$\eta = \begin{cases} 1.0 & \text{for normal weight concrete} \\ 0.8 & \text{for lightweight concrete} \end{cases}$$

A_{cv} is the average cross-sectional area per unit length of the beam of the concrete shear surface under consideration

A_{sv} is the cross-sectional area per unit length of the beam of the combined top and bottom reinforcement crossing the shear surface, see Fig. 3.52.

The longitudinal shear force per unit length, v, to be resisted, can be obtained from the spacing of the shear connectors

$$v = \frac{NQ}{s} \leq v_r$$

where N is the number of shear connectors in a group

Q is the shear connector resistance for positive (sagging) moments

s is the longitudinal spacing of the shear connectors.

This shear force may be reduced in accordance to the actual force transferred if more than the required amount of shear connectors is provided.

3.9.11 *Fire resistance design*

Slimflor beams have inherently good fire resistance because they are partially encased within the floor slab. A fire resistance of up to 60 minutes can generally be achieved without the need for applied fire protection. For longer periods of fire resistance, fire protection to the bottom plate is required.

The important parameter is the rate of heating of the RHS section which has been established by small-scale tests and by a major full-scale fire test carried out at TNO in the Netherlands.

It is not intended to review the fire engineering method or all of the existing test information in detail, but rather to concentrate on the implications for RHS Slimflor edge beams.

3.9.12 *General requirements of BS 5950: Part 8*

No specific requirements for Slimflor construction are presented in BS 5950: Part 8.[2] However, the strength reduction factors for structural steel and concrete at elevated temperatures are given. It is assumed that the strength reduction for steel corresponding to 2% strain is appropriate for slim floor beams at the large deformations experienced in fire.

Partial factors for loads are also defined in BS 5950: Part 8. A partial factor of 0.8 may be used for variable imposed loads and 1.0 for all permanent loads at the fire limit state. The ratio of the applied loads at the fire limit state to the load resistance of the member in normal conditions is generally in the range of 0.45–0.55 for practical design cases.

The minimum depth of composite slabs required to satisfy the insulation criterion of BS 476, the fire testing standard, is also given in BS 5950: Part 8.[2] Fire tests have confirmed that these are applicable to slabs constructed using SD225 decking. These values are presented in Table 3.5.

Table 3.5 Minimum slab depths for composite slabs with SD225 deep decking.

Fire resistance (minutes)	Slab topping depth (mm)		Total slab depths (mm)	
	NWC	LWC	NWC	LWC
30	60	50	285	275
60	70	60	295	285
90	80	70	305	295
120	95	80	320	305

The integrity criterion of BS 476 is satisfied by the steel decking beneath the slab and mesh reinforcement in the concrete topping. The moment resistance of a composite slab is established from first principles using the method described in BS 5950: Part 8, taking account of the reduced strength of all the elements in fire conditions.

3.9.13   *Additional fire protection*

The previous procedures apply where the bottom plate of the RHS is unprotected which is generally the case for 60 minutes of fire resistance. For longer periods of fire resistance, the recommended design procedure is to maintain a limiting temperature of not more than approximately 700 °C in the bottom plate by using an appropriate thickness of fire protection. At this temperature, the RHS Slimflor edge beam will achieve a load ratio of at least 0.6.

This protection thickness can be conservatively calculated using a section factor of $1/t_p$, where t_p is the plate thickness in metres. For example, when t_p is 15 mm, the section factor is 67 m^{-1}. Design tables for standard products are presented in the ASFPCM/SCI publication *Fire protection for structural steel in buildings*.[11]

Typically, a minimum of 12 mm of board or spray protection to the bottom plate is required for 90 minutes of fire resistance, increasing to 20 mm for 120 minutes of fire resistance. Intumescent coatings may also be used and there is sufficient test information to demonstrate their performance for up to 90 minutes of fire resistance. Some protection manufacturers will have more accurate data and may be able to offer appreciably reduced thicknesses of protection for all types of slim floor beam.

The side of the RHS should also be protected either by its attachment to the cladding material (such as blockwork or brickwork) or by a board or mineral wool material. Any gap between the RHS and cladding should be 'fire stopped' to prevent the passage of smoke and hot gases.

3.10   Composite slabs using deep decking (SD225)

The new deep decking profile, SD225 (see Fig. 3.53), has been the subject of extensive modifications to improve the spanning capability of the former 210 mm (CF210) deep profile. It is 225 mm deep and uses steel of grade Fe E350 and thickness of 1.25 mm. In addition, the deck configuration has been adapted to readily accommodate the ceiling and services without the need to make attachments into the concrete slab.

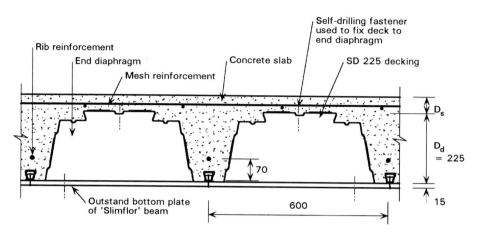

Figure 3.53   Typical cross-section through the floor slab using SD225 decking.

The three important design cases are:

• the ability of the steel decking to support the loads during
  construction without the need for temporary propping
• the imposed load resistance of the composite slab which is also
  influenced by the amount of bar reinforcement in the ribs
• the fire resistance of the composite slab as influenced by:
  – the minimum slab thickness for insulation purposes
  – the moment resistance due to the reinforcing bar in the ribs
    (assuming the steel decking to be ineffective in fire).

3.10.1   *Construction stage*

During construction, the steel decking is designed to support the self-
weight of the concrete and a construction load of 1.5 kN/m^2 (as specified in BS 5950: Part 4). Because the code does not envisage the use of
long-span slabs, the construction load for unpropped cases is taken as
1.5 kN/m^2 over the middle 3 m of the span and 0.75 kN/m^2 elsewhere
(as proposed in Eurocode 4: Part 1.1[13]). The loading arrangement in
unpropped construction is defined in Fig. 3.54. The partial factors in
BS 5950 are taken as 1.6 for construction loads and 1.4 for dead loads.
  Ponding of concrete, due to the deflection of the deck, is taken into
account in these construction loads provided the deflection of the deck
after construction does not exceed span/180. The deck shape is highly
stiffened to improve its bending resistance and by reducing the effects of
local buckling. Calculations and load tests carried out by British Steel
(Welsh Technology Centre), PMF Ltd and SCI have confirmed the
maximum distance between beams during the construction stage to be:

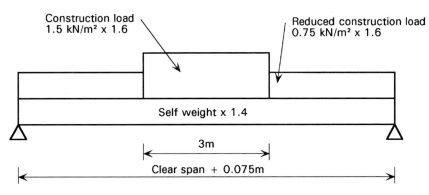

Construction load
1.5 kN/m² x 1.6

Reduced construction load
0.75 kN/m² x 1.6

Self weight x 1.4

3m

Clear span + 0.075m

Figure 3.54   Loading on deep decking at the construction stage.

- 6.0 m (unpropped) for normal weight concrete slab (295 mm overall depth)
- 6.5 m (unpropped) for lightweight concrete slab (285 mm overall depth).

The difference in spanning capabilities is due to the difference in slab weight, the slab depth also being influenced by the fire resistance requirements (see Section 3.9.12).

The composite (or shear bond) action of the slab with the steel decking is achieved by the use of vertical indentations in the webs of the decking. Also, the 'dovetails' positioned in the crest and trough will enhance the composite action of the slab. Also, additional reinforcing bars are necessary to provide fire resistance. However, these bars may also be included in the 'normal' design of the slab. It is generally found that the load-carrying capacity of the slab exceeds that required in most applications provided spans are less than 8 m. Engineers should always consult the suppliers of the decking (Precision Metal Forming Limited) for more detailed information and load–span tables.

When deep decking spans normal to the beam, end diaphragms are connected to the bottom flange plate or flange of the ASB which effectively provides a closure for the concrete. The steel deck is placed on these diaphragms and connected using shot fired pins (alternatively, pneumatic fasteners) to the projecting bottom plate or flange of the section. The in situ concrete is then poured or pumped onto the deck and levelled to a final finish.

For the ASB section, cut-outs are provided at the ends of the decking to facilitate placing of the concrete around the ASB section. Also, the cut-outs enable easier deck handling in the construction stage.

When the decking is propped during construction, the spacing of the props is determined by combined bending and shear in the hogging (negative) moment region of the decking over the line of props. It is recommended that the spacing between the props should be relatively

Figure 3.55   Deck testing using the vacuum rig.

Figure 3.56   Deck testing using applied point loads.

close, and a timber bearer should be used to avoid local indentation of the deck ribs at these points. The props should only be removed when the concrete has achieved at least 75% of its design strength.

Figures 3.55 and 3.56 show the deck testing arrangements using the vacuum rig and by point load testing. Note, the deck troughs (Fig. 3.56) are filled with sand to simulate the wet concrete condition in the construction stage.

3.10.2 *Section properties*

The section properties of deep deck profiles are enhanced by use of longitudinal stiffeners and transverse ribs. However, only a proportion of the cross-section is fully effective in bending. The SD225 deep deck profile is considerably more effective in bending than the CF210 profile, largely because of the re-entrant portion in the top flange (see Fig. 3.53).

The elastic bending resistance of deep decks is established by tests, because design to BS 5950: Part 6 is not directly appropriate to such highly stiffened deep profiles. The middle portions of the web and upper flange are largely ineffective due to the influence of local buckling; the effective section of the SD225 profile is illustrated in Fig. 3.57. The same elastic properties are used to determine the deflection of the decking after construction.

The properties of the CF210 and SD225 profiles (as justified by calculations and testing) are presented in Table 3.6. These properties are established for the standard steel thickness of 1.25 mm and yield strength of 350 N/mm^2.

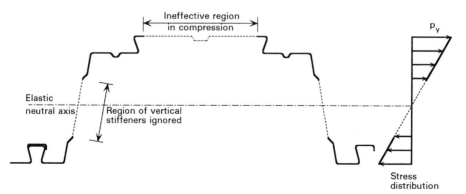

Figure 3.57   Effective section of the SD225 deck.

Table 3.6   Section properties of deep deck profiles.

Deck profile	Section modulus (cm^3/m width)	Second moment of area (cm^4/m)
CF210	66.3	855
SD225	75.0	950

$t = 1.25\,\text{mm}$ $p_y = 350\,\text{N/mm}^2$

3.10.3 *Composite slab design*

The design of composite slabs using deep decking follows the conventional principles established for shallow deck profiles. This book does not consider the design and testing requirements of composite slabs in detail which are otherwise covered by the provisions of BS 5950: Part 4.

Generally, it is the construction condition that controls the design of the slab. The SD225 deep deck profile has been designed to achieve optimum bending properties for this deck height. In all cases, the effective span of the decking or slab is taken as to the centres of bearing, i.e. is the clear span + 75 mm.

Minimum slab depths

The minimum slab depth for insulation purposes in fire determines the slab thickness and the self-weight of the slab, and hence the maximum span capabilities of the steel decking. Lightweight concrete is preferred in composite slabs because of its better insulating properties and lower density (approximately 1900 kg/m^3, or 75% of that of a normal weight concrete slab) which leads to thinner and longer-span slabs.

The minimum slab depth is also controlled by the depth of concrete over the top of the beam required to permit placing of the mesh reinforcement. It is considered that 30 mm cover is a sensible minimum, making a total slab depth of 290–315 mm depending on the ASB section size. This slab depth satisfies the insulation requirement for 60 minutes of fire resistance for both normal and lightweight concrete. However, the 300 ASB section may be designed to be level with the top of the slab for 30/60 minutes of fire resistance and achieves the required minimum slab depths.

The minimum slab depths (excluding the bottom flange thickness) are presented in Table 3.7.

It is difficult to give precise guidance for slab depths dealing with Slimflor beams and RHS edge beams due to the wide range of beam

Table 3.7   Minimum slab depths for use of ASB sections with SD225 decking.

ASB section	Fire resistance			
	60 min		90 min	
	LWC	NWC	LWC	NWC
280 ASB 100	290	295	295	305
280 ASB 136	295[b]	295[b]	295[b]	305
300 ASB 153	286[a] 315[b]	— 315[b]	— 315[b]	— 315[b]

All slab depths exclude the bottom flange of the ASB section.
All slab depths include 30 mm minimum top cover to the ASB section, except [a] which is flush with the top of the section.
[b] All dimensions are rounded down by 1 mm.

depths available. Also, the form of construction has an influence on the beam depth. Composite beams require an additional depth of concrete above the top surface of the beam to facilitate the use of shear stud connectors.

Composite action

Composite action in composite slabs occurs partly due to natural bond between the steel and concrete, but is mainly due to the mechanical interlock created by the embossments in the steel decking. The degree of this shear-bond action is established by standard load tests on slabs carried out to BS 5950: Part 4. The decking is tested in its as-delivered condition. The design bending resistance of the composite slab is based on empirical m and k parameters which are calculated from these tests using a regression line analysis as a function of the slab span and depth. The degree shear bond resistance is further reduced by a partial factor of 1.25.

Tests have shown that the bending resistance of the slab may be increased by the use of bar reinforcement in the ribs. This moment resistance is calculated from that of the equivalent reinforced concrete slab and is added directly to the resistance due to shear-bond action of the composite slab as the two forms of failure are 'ductile' in deep deck slabs. This mechanism is illustrated in Fig. 3.58. The bar reinforcement is also effective in fire conditions as it compensates for the loss in strength of the steel decking which is exposed directly to the fire. For most applications the centre of the bars is positioned at 70 mm above the soffit of the decking in all cases.

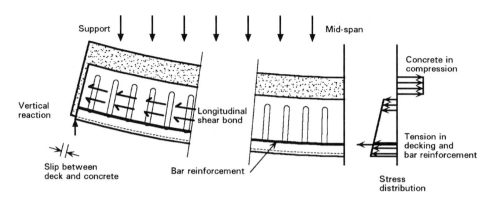

Figure 3.58   Action of composite slab with reinforcement in ribs.

The loads applied to the composite slab depend on whether or not the slab is propped. In the propped case, the loads due to removal of the props mean that the total factored load is applied to the composite slab. In the unpropped case, only the factored total load less the self-weight of the slab is considered in the design of the composite slab.

3.10.4 *Installation notes*

Decking

The decking must be positively fixed to the supporting structure. This is to ensure that movement of the decking during construction is kept to a minimum and also to prevent excessive deflection during the pouring of the concrete.

The required fixing frequency of main fasteners is two per trough at all support positions. When fixing onto a steelwork support structure, heavy-duty shot-fired pins or pneumatic fasteners are to be used. The steelwork must provide adequate support for the deck around columns and openings. In the case of other support systems, such as brickwork, blockwork and concrete, the profiled decking must be fixed using adequate masonry fixings. Also, the deck crest (in all cases) is fixed to the 50 mm returns of the diaphragms with self-drilling fasteners at a frequency of two fixings per deck pitch. Note that damaged deck *must* be replaced.

End diaphragms

End diaphragms are provided with the decking in order to contain the concrete at the deck ends during the concrete pour. These diaphragms are manufactured from 1.6 mm gauge galvanised steel and are 1800 mm long (3 × 600 mm). They have to be fixed to the support structure (bottom flange) before the decking is laid. The decking is then positioned over the diaphragms and fixed to the support structure. The diaphragms are connected to the support structure with a minimum of two heavy-duty shot-fired pins or self-drilling/tapping fasteners. These diaphragms also provide exact alignment of the decking during construction and additional stiffness to prevent web buckling of the deck.

Rib reinforcement and mesh placement

The deck design requires the inclusion of a single reinforcing bar in each profile rib. The bar size can vary from 10 mm to 32 mm diameter and is positioned with the centre of the bar 70 mm above the profile bottom flange.

Over the deep decking, standard reinforcing mesh, i.e. A142, A193 and A252 designation, can be used which is positioned towards the top of the slab. The top cover to the reinforcement mesh should be a minimum of 15 mm and a maximum of 45 mm, with minimum laps of 300 mm for A142 mesh and 400 mm for A193 and A252 mesh.

Casting concrete

Before commencing the concrete pour, the decking must be cleared of all dirt and grease which could otherwise influence the performance of the hardened slab adversely. The oil left on the decking from the roll-forming process does not have to be removed.

Care should be taken to avoid heaping of concrete in any area during the casting sequence. The concrete should be discharged in a controlled manner from a level not exceeding knee height (0.5 m).

Edge trim

Edge trim is used to retain the wet concrete slab and to determine the thickness of slab at the building perimeter. The edge trim is normally supplied in 3 m lengths, varying from 1.2 mm to 2.0 mm in thickness. Edge trim is usually shot-fired to the steel support structure and the top of trim is connected to the decking with restraint straps at approximately 350 mm centres using either pop-rivets or self-drilling fasteners.

Chapter 4
Other Forms of Composite Construction

This section provides information for longer-span solutions. The most popular forms of construction are:

(1) Composite trusses
(2) Fabricated composite beams
(3) Stub girders
(4) The parallel beam approach
(5) Cellular beams

The following sections briefly explain the relevant design criteria. (For further design information relating to the above topics, see references 17–21.)

4.1 Composite trusses

4.1.1 Introduction

Composite construction in buildings is well established and there is often a demand for longer column-free spans in buildings. Conventional steel frames with concrete or composite slabs may be used but often the size of long-span beams is such that the floor zone is excessively deep. There is also a need to incorporate a high degree of servicing in modern buildings and coupled with this is the need to minimise floor zones where building heights are restricted, or to reduce cladding costs.

Various design solutions are feasible but there are two basic options: either the structure and services are integrated within the same horizontal zone, or the structural zone is minimised so that the services are passed beneath. These solutions are briefly described in the following section.

For long-span structures, integration of the structure and services is the optimum solution. The economics of building construction are such that a small increase in the cost of the structure is often not

Figure 4.1   Example of a composite truss.

significant in terms of the overall building cost, provided there are additional benefits in flexibility of servicing and internal planning.

One of the best potential solutions for beam spans in the region of 12–20 m is the 'composite truss' as illustrated in Fig. 4.1. This form of construction is popular in North America and is often used for long-span beams where the truss occupies the entire ceiling-to-floor zone. This provides the maximum zone for services to pass through the voids created between the chord and bracing members.

Trusses are usually of a standard 'Warren truss' form with the node points at the top chord arranged to coincide with any point load positions. Often a central 'Vierendeel' panel without bracing members is introduced to provide a greater zone for services. The truss is usually fabricated from angle sections as the bracing members and T-sections (cut from Universal Column Sections) as the chords. Composite action provides a moderate increase in strength, but importantly, in long-span applications, it greatly improves the stiffness of the steel truss.

4.1.2   *Composite truss systems*

Range of application

Composite trusses combine the efficiency of the truss form with the speed and cost savings of composite construction. The truss system concentrates material at the structurally most efficient locations for force transfer, hence, for a given span, the composite truss theoretically has the least steel weight of any steel framing system. In addition, the relatively open area of the truss system can freely accommodate a substantial amount of building services. Against these advantages are disadvantages of additional fabrication and fire-proofing costs of trusses relative to more conventional framing systems.

Experience has shown that trusses are economically viable for spans greater than 12 m. For spans between 12 m and 15 m, the cost competitiveness of trusses is determined by floor-to-floor height limitations and the services configuration and size. Where spans of 18 m or longer are required, composite trusses are generally the most economic structural system.

Framing arrangements

Two basic framing arrangements of composite trusses are normally used. In the first, the trusses form the secondary framing elements and support the decking directly. In the other, trusses form the primary framing elements that support secondary beams, these in turn support the decking. Generally 'composite trusses' are defined as those used as secondary beams and 'composite truss girders' are those used as primary beams (see Fig. 4.2).

A number of variations of these basic framing arrangements are possible, for example, where spans are large in both framing directions, composite truss girders can be utilised to support secondary composite trusses.

The spacing between composite trusses is governed by the maximum span of the slab or deck. To economise on the number of members required, this spacing should be as large as possible. For the shallow decks currently available, and the slab fire ratings required, the maximum truss spacing is approximately 4.0 m, with 3 m spacing being the most common. With the introduction of deeper decks, the spans could increase to 6.0 m, which is a more efficient truss spacing.

Truss configurations

A large number of truss configurations are worthy of consideration (see Fig. 4.3). However, the Pratt (or N-truss) and Warren truss configurations (A and B) are the most common. Pratt trusses, although theoretically the most efficient truss configuration, have

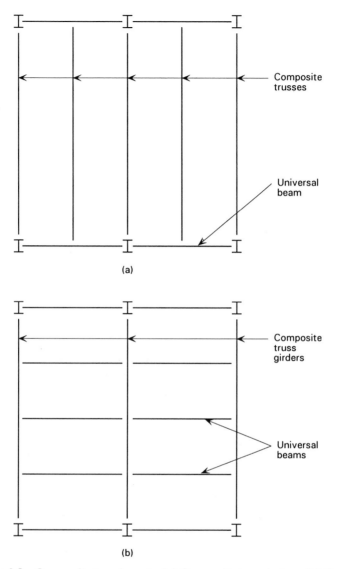

Figure 4.2 Composite truss layouts. (a) Composite truss system. (b) Composite truss girder system.

limited usefulness for typical floor framing. The additional components increase fabrication costs and the relative small free area between the diagonals greatly reduces flexibility for service sizes and locations.

A conventional Warren truss configuration limits service duct sizes to those that will fit between the diagonal bracing members. However, most truss applications (C–G) will permit the incorporation of Vierendeel panels without bracing members which greatly increase

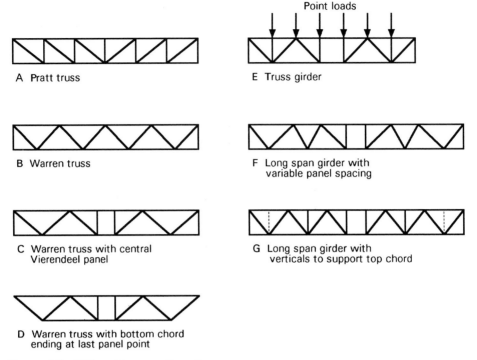

Point loads

A Pratt truss

B Warren truss

C Warren truss with central
   Vierendeel panel

D Warren truss with bottom chord
   ending at last panel point

E Truss girder

F Long span girder with
   variable panel spacing

G Long span girder with
   verticals to support top chord

Figure 4.3 Different truss configurations.

the zone for services. In order to maximise the size of the openings and
to minimise reinforcing of the truss chords, the Vierendeel panel
should be located at mid-span.

The bottom chord can either extend to the support or terminate at
the last panel point, as in configuration D of Fig. 4.3. Generally, where
trusses are used as secondary members, the chord can terminate before
the support. Where the truss acts as a primary beam, or supports
heavy point loads, it is recommended that the chord extends to the
support to provide improved resistance to 'flange tripping'.

Vertical members can be introduced in Warren trusses, as in
configurations E and G of Fig. 4.3, in order to reduce the span of the
top chord and hence minimise the top chord size. This is especially
advantageous in the first panel either side of the Vierendeel panel
where top chord bending and axial forces are the highest. Varying the
panel spacing, as in configuration F of Fig. 4.3, can achieve the same
benefit. However, the disadvantage is the fabrication of varying
bracing member lengths.

The centroidal axes of the compression and tension web members
should ideally meet at the same node point on the chord axis.
However, in the composite case, the effective node point is within the

slab and therefore the chords can be 'separated' slightly. If the chord members are rectangular hollow sections (RHS), the bottom chord web joint can accommodate a slight eccentricity. However, for T-section chords, the additional moments that are induced can influence the required section size. Therefore, for most applications, concentric joints should be used, with the connection centre lines of the web members separated only as necessary to simplify fabrication.

The Warren truss with a central Vierendeel panel and a discontinuous bottom chord (Fig. 4.3, truss D) will usually be the most efficient configuration. However, the number of bracing members is important as regards the overall economy of design and use of trusses of this type. Three cases are shown in Fig. 4.4.

Type A has advantages of:

(1) The top chord has a shorter span between nodes permitting use of a smaller member.
(2) The increased angle of incidence of the bracing members reduces their forces and their length, permitting use of smaller members and gussetless construction.

Type B has advantages of:

(1) Fewer members have reduced fabrication costs (although sharp intersections are more difficult to weld).
(2) Greater space for servicing.
(3) Fewer members to fire protect.

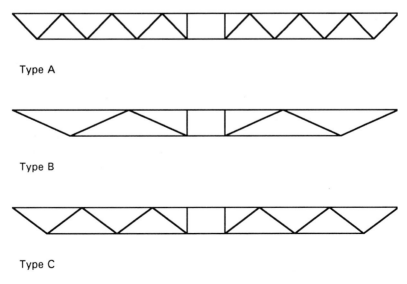

Type A

Type B

Type C

Figure 4.4 Examples of different Warren truss panel spacings.

Type C is probably a good compromise among all the above criteria. For a truss girder, the location of the top nodes should align with any secondary beams.

4.1.3 *Truss member types*

Any structural section can be used for both the chord members and the web members of the truss. However, T shapes or rectangular hollow sections are most commonly used as chord members. Angles, either single or double, are usually used for web members in association with T and angle chords. RHS or circular hollow section web members (S275) are commonly used with RHS chords (S355). If a small diameter tube was used for the diagonal member and a large RHS was used for the chord of the truss, then the punching effect on the wall of the chord section could have serious consequences.

The use of different grades of steel produces a joint where the interesting members are of similar size. This has the advantage of reducing the possibility of local failure in this vicinity. Figure 4.5 shows some of the possible configurations.

Chord members

The selection of either RHS or T-sections for chord members provides different advantages. RHS sections are more structurally efficient and therefore are potentially shallower than T-sections. However, RHS sections are more costly but lighter in weight and provide an aesthetic appearance. The wall thickness may be governed by requirements for welding of the shear connectors (see section 3.9.5). Tee sections cut from UC sections are usually the optimum choice for all the chord members. The stems of the T-sections should be sufficiently deep to permit easy welding of the bracing members.

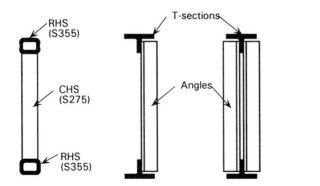

Figure 4.5 Truss member types.

Web members

Single- or double-angle web members allow use of gussetless welded connections which minimise fabrication costs. Other member types either require gussets or more complex welding, e.g. CHS to CHS. Single-angle web members are often more cost effective than double angles given that for a slight increase in material weight, the number of welding operations can be reduced. Single angles should be placed on alternate sides of the top and bottom chords. If not, the chords must be designed for the additional eccentricity caused by the transverse bending effect which can have a significant effect on member size. For 12–15 m spans, single angles can usually be used in this manner. However, longer-span or heavily-loaded trusses generally require double angles.

Trusses with a mixture of single- and double-angle web members can also be used. In areas of high shear (near the ends of the span) and/or high bending forces (adjacent to the Vierendeel panels), use of double angles may minimise material costs. In other areas use of single angles minimises fabrication costs.

Double-angle web members can either have spacers and hence be designed using their battened properties, or be designed as two independent single angles. Although battened angles usually require less material, the fabrication savings derived from not using spacers usually outweighs the additional material cost. Therefore, for trusses in the mid-span range (12–15 m), use of non-battened angles should be considered.

Eccentricities at joints

The centroidal axes of the compression and tension web members are not necessarily required to meet at the same point on the chord axes. In fact, trusses with eccentric web–chord joints can be as efficient as those with concentric joints. In the composite case, the nodal point is in the slab, so the web to top chord connections may be 'separated' to eliminate joint eccentricity in the composite truss. Often, the bottom chord–web joint can also be 'separated' without an increase in steel weight. Although eccentric joints require that local moments be designed for, several advantages also accrue. Eccentric joints provide additional space for welding, hence reducing fabrication complexity. In addition, the reduced length of the compression chord provides enhanced buckling and bending resistance which partly compensates for the additional moments generated by the joint eccentricity.

Therefore, where the bracing members are placed as double angles, some slight joint eccentricity is often an advantage and the angles can be designed for the local moments. However, for single-angle bracings it is recommended that joint eccentricity is minimised. This is not difficult to achieve as they are to be placed alternately on either side of the chord section.

4.1.4 *Hand analysis of composite trusses*

Hand analysis can be carried out both for initial and final design. However, many designers use hand analysis to obtain trial member sizes which are then input into a 'frame' analysis. The following sections describe the main features of the hand analysis. Cross-reference to BS 5950: Part 1 and Part 3 is made, where appropriate.

Initial design

The depth of the steel truss is selected to satisfy the span/depth criteria of a simply supported structure (typically span/15 to span/20) and to accommodate the architectural limitations on the allowable ceiling depth. Sufficient depth allowance (typically 150–250 mm) should be made below the bottom chord member so as to accommodate truss deflections, fire protection and the lighting and ceiling systems.

The depth of the truss is often governed by the clear area required for routing of the major services through the truss. To maximise this space, the slopes of the bracing diagonals should be selected to be 45° or less to the horizontal. A slope of around 30°, creating a panel width to truss depth ratio of 3:1, has been found to be the most efficient proportion. The larger panel sizes have fewer, even if slightly heavier, members, hence minimising fabrication and fireproofing costs. Similarly the size of the Vierendeel panel should be such as to accommodate the major service duct, with the maximum width of the opening limited to approximately 1.5 times the depth of the truss.

The sizing of the top chord is influenced by the following criteria:

(1) Its ability to span between the braced nodes and to support the loads during construction.

(2) To provide bearing support for the decking, and stablility during erection. A minimum flange width of 120 mm is usually acceptable.

(3) Where through-deck welding is to be used, the minimum flange thickness is 8 mm for use with 19 mm diameter studs.

(4) Resistance to local bending at Vierendeel openings.

(5) Welding of the bracing members.

Table 4.1   Section classification of T-sections cut from UC sections.

T-section width × depth × weight	Grade S275	Grade S355
203 × 102 × 43[a]	C	SC
254 × 127 × 66[a]	C	SC
254 × 127 × 84[a]	P	C
305 × 152 × 99	C	SC
305 × 152 × 120	P	C
305 × 152 × 142	P	P

All T-sections are semi-compact (SC) except for the plastic (P) and compact (C) sections
[a] May not be sufficiently deep to accommodate the weld length of the bracing members.

A T-section cut from a UC is usually detailed, the size of the section depending on the conditions in mid-span. Members that are 'plastic' or 'compact' in bending or compression according to Table 7 of BS 5950: Part 1 can be used most efficiently in design. However, relatively few T-sections achieve this limit (see Table 4.1). Chords cut from UB sections are usually 'semi-compact' or 'slender' when considering their stems in compression. Slender sections are inefficient because only a portion of their depth is effective in compression and bending.

Moment capacity of the composite truss

In limit state design, the moment capacity of the composite truss should exceed the total factored moment applied to the beam. However, additional checks are required on the steel truss in the construction condition (see construction condition, page 126). This is usually only important when considering the design of the top chord and the Vierendeel panel.

The moment capacity of a steel truss system at the point of maximum moment is determined by compression in the top chord and tension in the bottom chord. In a composite truss, the compression force may be considered to be resisted by the concrete or composite slab with a consequent increase in the lever arm from the bottom chord to the point of compression in the slab. The different behaviour of composite and steel trusses is illustrated in Fig. 4.6.

The important parameters in this analysis are the tensile resistance of the bottom chord, R_b, and the compressive resistance of the concrete slab, R_c. The contribution of the top steel chord is ignored because of concern about the amount of strain in the bottom chord necessary before the full tensile action of the top chord is developed.

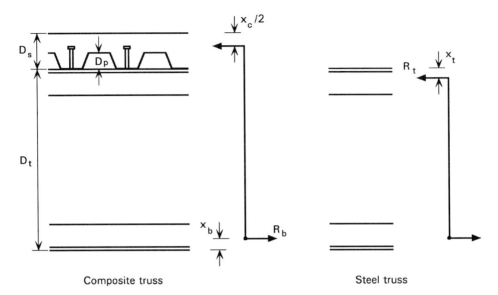

Figure 4.6   Moment capacity of composite and steel trusses.

The resistance of the bottom chord is given by:

$$R_b = A_b p_y$$

where A_b is the cross-sectional area of the bottom chord
p_y is the design strength of steel.

The compressive resistance of the slab (above the deck) is R_c which acts at a stress of $0.45 f_{cu}$. Hence,

$$R_c = 0.45 f_{cu} B_e (D_s - D_p)$$

where f_{cu} is the cube strength of concrete
B_e is the effective breadth of the concrete slab
D_s is the overall slab depth
D_p is the depth of the deck profile.

The effective breadth, B_e, is defined in BS 5950: Part 3 as (beam span)/4 for internal beams but not exceeding the beam spacing, b, for secondary beams or $0.8b$ for primary beams.
In most cases $R_b < R_c$ and so the moment capacity, M_c, of the composite truss is given by the tensile action of the bottom chord multiplied by the lever arm to the point of compression in the slab. Hence

$$M_c = R_b (D_t + D_s - 0.5 x_c - x_b)$$

where $x_c = (D_s - D_p)\dfrac{R_s}{R_c}$

D_t is the overall depth of steel truss

x_b is the depth of elastic centroid of the bottom chord above the bottom of the truss.

The increase in moment capacity of a composite truss is rarely more than 20–30% in comparison to the parent steel truss. There are, however, other benefits of composite action in terms of local moment transfer at Vierendeel openings and the increase in the overall stiffness of the truss.

Shear capacity of composite truss

The shear capacity of a truss can be evaluated from first principles by considering the component forces in the bracing members (Fig. 4.7). All connections are assumed to be pinned at this stage. In the conventional Warren truss, the outer bracing members are in tension. If the bracing members are orientated at an angle θ to the horizontal, their maximum tensile force is:

$$F_t = \frac{\text{reaction}}{\sin \theta}$$

It follows that the bracing force increases significantly as θ reduces.

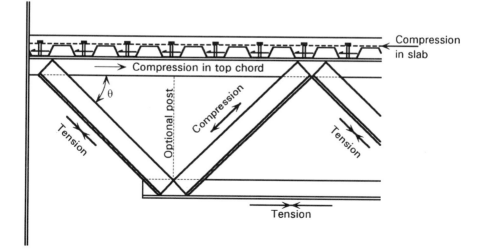

Figure 4.7 Forces in bracing members.

This tensile force is resisted by a compression force in the next bracing member remote from the support. If no vertical post is provided, then the tension and compression forces are equal. If a vertical post is provided (although this reduces the service zone), the compression force is reduced as a result of the local force transferred from the slab via the post.

A compression force is developed in the steel top chord due to the component of force transferred via the bracing members. However, in a composite truss, this force is then transferred uniformly to the slab by the shear connectors attached to the top chord.

The bracing forces may be calculated at all points along the truss by observing equilibrium at the nodes. For simple truss configurations, Bow's notation may be used (graphical method). The size of the bracing members may be reduced, if desired, towards the low-shear (i.e. mid-span) zones.

The tensile resistance of the members may be fully developed, except in bolted construction where allowance should be made for the loss in area of the bolt holes. Bolted construction is not recommended for long-span trusses (except where close tolerance bolts are used) because of the cumulative effects of slip on deflections.

The compressive resistance of the members depends on their slenderness between the nodes. For single angles, the slenderness is based on the u–u or v–v axes of the sections as defined in BS 5950: Part 1, Table 28. For double angles, connected periodically along their length, the slenderness is usually based on the x–x or y–y axes. Knowing the slenderness of the members, their compressive strength is obtained from Table 27 of BS 5950: Part 1.

Local moments at braced nodes in the bottom chord are usually ignored at this stage. Where there is an eccentricity in the projected centroids of the members, the eccentric force gives rise to an overall bending effect at the node which is resisted by the members broadly in proportion to their stiffness. The members may be checked for combined bending and axial effects, as in BS 5950: Part 1.

The treatment of the nodes in the top chord connected to the composite slab is interesting. In this case, the bracing members may be separated slightly as their projected centroids need not align with the steel top chord but with the centre of the concrete slab. Any further eccentricity may be treated as a local moment in the slab.

Vierendeel panels

Where bracing members in the low-shear zone (in mid-span) are removed and vertical members are introduced, a so-called Vierendeel

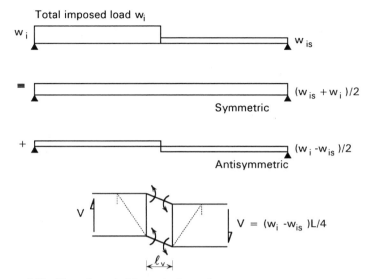

Figure 4.8 Shear force in Vierendeel panel.

panel is created. In a non-composite truss, the local moments that are created are resisted by the steel chords. However, in a composite truss, it is possible to make use of the moment capacity of the top chord acting compositely with the concrete slab.

In order to take account of the possibility of shear transfer through the ideally 'zero shear' zones, it is proposed that the following 'worst' load case is considered; full factored imposed load on half of the span and no imposed load on the other half. The imposed dead loads are applied uniformly throughout. The net effect is to cause a shear force in the Vierendeel panel equal to 25% of the total factored imposed load on the entire span of the truss. The tension in the bottom chord is also reduced. This load case is illustrated in Fig. 4.8. The Vierendeel moment transfer is $V\ell_v$, where V is the shear force (as determined above) and ℓ_v is the length of the panel opening. This is resisted by moments in the chords.

The distribution of moments around the Vierendeel panel is complex and a significant component of resistance is the composite action developed between the top chord and the slab. Neglecting this action will lead to very conservative design and hence to difficulty in justifying the use of wide openings that are needed in practice.

An additional criterion is that the chord members and the concrete slab can resist the shear force, V. The shear resistance of the steel chords can be calculated from BS 5950: Part 1. The shear resistance of the concrete slab is ignored at this stage.

Construction condition

In the construction condition, the steel truss is designed to support the weight of wet concrete and a construction load equivalent to a uniform load over the supported area of $0.5 \, kN/m^2$, or a single point load of 4 kN. Local bending may influence the minimum size of the top chord section that can be used.

The moment capacity of the section may be determined from first principles as illustrated in Fig. 4.6. If, as generally is the case, the size of the top chord is less than that of the bottom chord, the moment capacity of the truss is given by

$$M_c = R_t(D_t - x_b - x_t)$$

where R_t is the compressive resistance of the top chord
x_t is the depth of elastic centroid of the top chord below the top of the truss.

The other terms are as defined previously in the section on moment capacity of the composite truss, page 121.

This moment capacity should exceed the factored moment on the section in the construction condition (using the load factors in BS 5950: Part 1).

No checks are required on the capacity of the bracing members at this stage, as the member design is controlled by the ultimate load condition.

In determining the maximum shear force across the Vierendeel panel opening, it should be assumed that half the span could be fully loaded during construction. Therefore, the shear force is equivalent to 25% of the total load due to self-weight and construction loads on the entire span.

The chords may be checked for the combination of tension or compression, and local moment. Each chord may be assumed to resist a shear force in proportion to its stiffness and the local moment is equivalent to the shear force times half the opening width. This approach is usually sufficiently accurate at the construction stage.

Longitudinal shear transfer in composite trusses

The force to be transferred by the shear connectors between the points of zero and maximum moment is R_b (when $R_b < R_c$). Full shear connection is achieved when the force transferred by the shear connectors in the half-span exceeds R_b. This force is equivalent to the number of shear connectors in the half-span times their design strength, as obtained from BS 5950: Part 3. It is usual practice to distribute the shear connectors uniformly along the truss, or in proportion to the shear force diagram in point-loaded beams. Partial shear connection principles can be used to reduce the number of shear connectors, but this is restricted to spans of less than 16 m.

An additional requirement is that sufficient transverse reinforcement, i.e. perpendicular to the axis of the truss, is placed in the slab so as to permit a smooth transfer of force from the shear connectors into the concrete. The requirements for transverse reinforcement are given in Chapter 3.

Deflection of composite and steel truss

The second moment of area, I_c, of a composite truss may be evaluated by reducing the concrete area to an equivalent steel area. Hence the truss becomes equivalent to two concentrated blocks of steel of area separated by the distance between the mid-depth of the slab and the bottom chord. This leads to the following expression

$$I_c = \frac{A_b A_c / \alpha_e}{(A_b + A_c / \alpha_e)} (D_t + (D_s + D_p)/2 - x_b)^2$$

where A_c is the cross-sectional area of concrete in the effective breadth of slab $(A_c = (D_s - D_p)B_e)$

α_e is the modular ratio between steel and concrete.

The other terms are as defined previously in the section on moment capacity of the composite truss, page 121. Conservatively, the area of the steel top chord is neglected as being small in comparison to A_c/α_e.

The modular ratio, α_e, is the ratio of the elastic moduli of steel and concrete. BS 5950: Part 3 recommends that α_e is taken as 10 for normal weight concrete and 15 for lightweight concrete when calculating imposed load deflections in office type buildings.

The imposed load deflection of a composite truss subject to uniform load w_i (per unit length) may therefore be calculated from the assumption that bending effects dominate. Hence

$$\delta_i = \frac{5 w_i L^4}{384 E I_c}$$

where E is the elastic modulus of steel $(= 205 \text{ kN/mm}^2)$

L is the span of the truss.

No further checks on serviceability performance are needed, except for calculation of self-weight deflections. The second moment of area of the steel truss, I_t, is obtained by considering the separation of the bottom chord and the top chord of area A_t. Hence

$$I_t = \frac{A_b A_t}{(A_b + A_t)}(D_t - x_b - x_t)^2$$

Generally, I_t will be significantly smaller than I_c, suggesting that self-weight deflection may be of similar magnitude to the imposed load deflections.

4.1.5 *Typical detailing*

Web and chord connections

The welded connections between the angle bracing members and the T-chords are illustrated in Fig. 4.9. The lines of force align to avoid eccentricity effects in the chords.

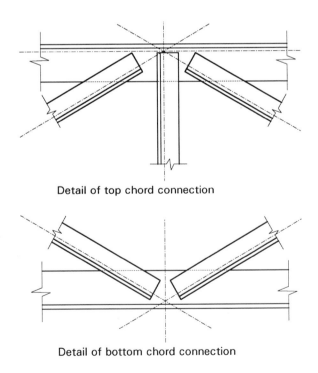

Detail of top chord connection

Detail of bottom chord connection

Figure 4.9   Chord/web connections.

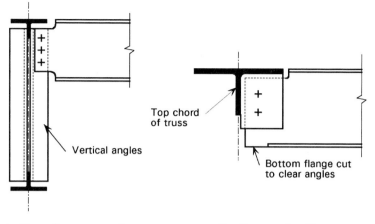

Top chord
of truss

Vertical angles

Bottom flange cut
to clear angles

Typical beam/truss connection
for heavy reactions

Detail at top chord/beam connection
for small reactions

Figure 4.10   Secondary framing connections.

Secondary framing

Attachment of secondary beams may be of the form illustrated in
Fig. 4.10, the choice often depending on the size of the members to be
connected.

End connections

End supports to composite trusses may be of the form illustrated in
Fig. 4.11 (page 130). For primary beams it is necessary to continue the
bottom chord to attach to the column to avoid flange tripping.

4.2   Fabricated composite beams

4.2.1   General description of structural arrangements and forms using fabricated sections

Figure 4.12 shows three typical floor arrangements for a one-bay long-
span structure. Wider, multi-bay buildings would simply be repetitious
of these single-bay arrangements, although the 6 m column spacing
that is shown along the building would be likely to increase for the
internal columns. Figure 4.12(a) shows the fabricated beams acting
as the primary beams, supporting light, rolled-section, composite,
secondary beams at 2.4–3.6 m centres (Type A). In Figure 4.12(b) the
fabricated sections are themselves placed at 2.4–3.6 m centres and are
supported directly by the columns or by composite beams (Type B).

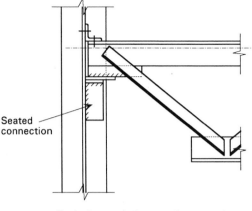

Seated
connection

Typical truss /column web connection

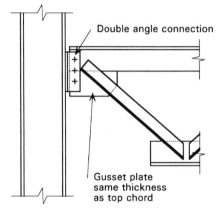

Double angle connection

Gusset plate
same thickness
as top chord

Typical truss /column flange connection

Figure 4.11   End connections for trusses.

In multi-bay schemes the internal beams would be replaced by primary beams on internal column lines. Figure 4.12(c) is only applicable to one-bay structures, with beams on the centre lines of all the columns (Type C).

The choice of arrangement will depend on the overall structural form. Type C would only be used if columns were required at centres of between 2.4 m and 3.6 m to support the building envelope.

Where the column spacing is greater than 3.6 m along the building, some form of grillage is required if conventional composite decking is used. The choice between Type A and Type B is not clear cut. For conventional construction, Type B would generally be favoured because it reduces the spans of the primary beams. However, layout

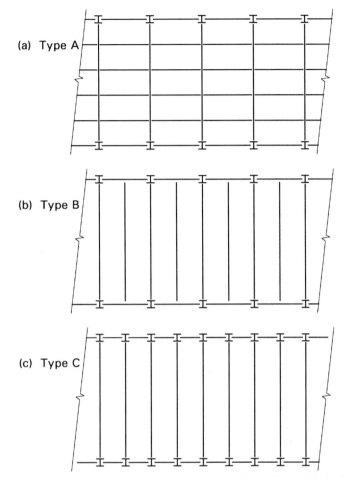

(a) Type A

(b) Type B

(c) Type C

Figure 4.12 Typical floor beam arrangements for a 15 m wide building. (a) Type A, (b) Type B, (c) Type C.

B does have a much greater number of fabricated sections which are inherently more expensive per tonne than rolled sections. In addition, the lightly loaded fabricated sections of Type B are likely to be less efficient than the heavier fabricated sections of Type A. For example, the webs of the former may be governed by minimum thickness criteria. Even if that is not the case, their greater slenderness will reduce strength. Conversely, both the piece count and number of connections in Type A are greater than Type B, thus increasing erection and fabrication costs for the former.

In many cases, service requirements will dictate choice of scheme. For example, Type A can accommodate much larger air-conditioning distribution units between primary beams than Type B without increasing the floor zone.

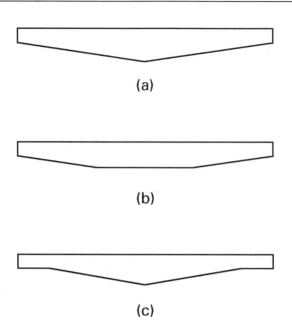

Figure 4.13 Shapes of fabricated beams for composite commercial buildings. (a) Straight taper. (b) Semi-taper. (c) Cranked taper.

Figure 4.13 shows a range of profiles for tapered beams. The most
straightforward is the linear taper (see Fig. 4.13(a)) which simply
comes down to a point at mid-span. Where depth of construction is at
a premium, as is usually the case, it is possible to keep to the same profile at the ends of the girder but maintain a constant depth over the
central portion of the beam, giving the semi-tapered shape shown in
Fig. 4.13(b). Where the triangular spaces beneath the profiles in (a) and
(b) are insufficient to accommodate the services, then the cranked taper
(Fig. 4.13(c)) can be adopted, with constant depth sections at the ends
of the beam. Choice of profile will primarily be dictated by services
requirements. Span-to-depth ratios may vary between 15 and 25; the
depth at the supports is usually reduced to half that at mid-span.

4.2.2 Accommodation of services

The principal reason for using tapered fabricated sections is that large
service ducts can be accommodated readily with the same depth as the
steelwork. However, because of the shape of the bending moment
diagram and the restrictions it imposes on beam depth, these ducts can
only be accommodated near the ends of the beams. Where this is not
acceptable to the services engineer, a different structural scheme

should be adopted. It is of course possible to penetrate the beam web for secondary services but when the web is slender such openings are likely to require stiffening if they are a significant proportion of the beam depth.

One significant flexibility of this form of structure is that there is a reasonably 'loose fit' between the structure and the services. If ductwork cannot be entirely accommodated within the depth of the steel section, the introduction of small coffers in the ceiling can overcome the problem, as well as give an attractive architectural solution, as shown in Fig. 4.14. This could be a particular advantage when the building is refurbished and reserviced during its life.

Where the floor layout in Fig. 4.12(a) is adopted, it is possible to accommodate large air-conditioning distribution units in the zone between the primary beams, as shown in Fig. 4.15, because the secondary beams will be less than 305 mm deep.

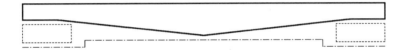

Figure 4.14   Use of coffered ceiling to accommodate large ducts.

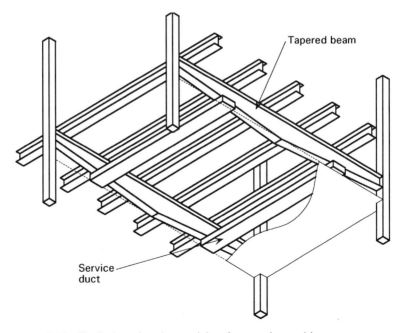

Figure 4.15   Typical services layout (view from underneath).

4.3 Stub girders

4.3.1 Introduction

The basic structural action of a stub girder is such that the resistance to applied moments is developed by tension in the lower chord and compression in the concrete or composite slab. The forces are transferred between these elements by 'stubs' or short sections of beam attached to the lower chord by welding or bolting, and by shear connectors to the concrete slab. In the orthogonal direction, secondary beams (sometimes referred to as purlins) achieve continuity by spanning over the bottom chord. The secondary beams are often designed with pin connections at the quarter span points.

In the basic stub girder system, the depth of the secondary beams is the same as the depth of the opening (see Fig. 4.16(a)). In this system the stubs are formed from the same section as that used for the secondary beams. There is therefore an optimum relationship between the size and span of the secondary beams (and hence the depth offered for servicing) and the span of the stub girder. Taking, for example, a column grid of 12 m square, the continuous secondary beams could be a 406 UB section. Assuming that the bottom chord is a heavy 254 UC section and the slab is 130 mm deep, it follows that the overall structural depth is approximately 800 mm, corresponding to a span : depth ratio of 15. A more optimum grid might be a 12 m × 15 m in which the stub girder spans the longer distance. Assuming the bottom chord is increased to a 305 UC section, the overall span : depth ratio increases to nearer to 18 which is more typical of 'regular' composite construction.

In the basic system, the stubs are sized so that the shear connectors needed to develop the appropriate force in the concrete are distributed at not less than the minimum spacing recommended in BS 5950: Part 3. This determines the length of the stubs and, consequently the maximum width of the openings available on either side of the stub. Clearly, the length of the stubs decreases as the force transferred decreases. This means that wider openings can be provided towards the middle of the span.

The bottom chord is designed to resist the combined tension, shear and moment developed under composite action. It is not usually sufficiently strong to resist loads developed during construction and, therefore, temporary props are required at normally the mid-span or third-span locations. The basic system can be modified, as shown in Fig. 4.16(b), to permit the use of unpropped construction by introducing a T-section as an upper chord which is designed to resist

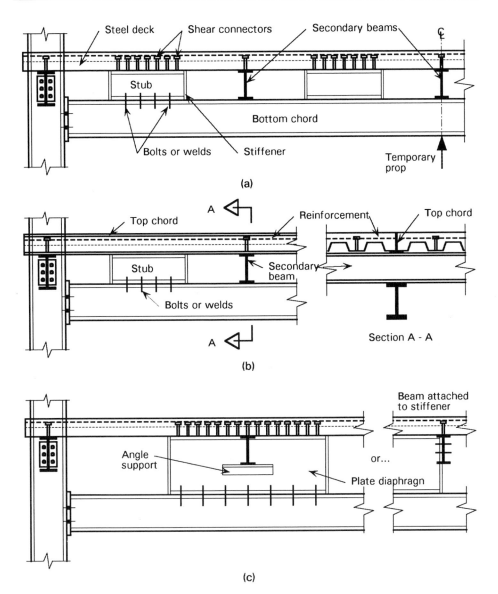

Figure 4.16 Different forms of stub girder. (a) Type A: opening depth to secondary beam depth, no top chord. (b) Type B: as Type A but with top chord to avoid temporary propping. (c) Type C: opening depth greater than secondary beam depth, no top chord.

compression when the stub-girder is subject to the self-weight of the floor slab (i.e. wet concrete) and other construction loads. This T-section is subsequently embedded in the slab. Holes can be drilled in the T-stalk so that reinforcing bars may be passed through and held in position, thus avoiding the need for shear connectors at the stub locations.

Another possible modification to the basic system is shown in Fig. 4.16(c). This addresses the common need for deeper opening zones than are obtainable for efficiently designed secondary beams. In this approach, deep diaphragm 'stubs' are fabricated from welded plate and the secondary beams can be attached to them by angles or web cleats. The location of the diaphragms can be different from that of conventional stubs (compare Figs 4.16(a) and (c)). However, in such a design the advantage of continuous secondary beams is lost unless holes are cut in the diaphragms and the beams passed through. This results in buildability problems, and therefore it would probably be more practical to design the secondary beams as simply-supported composite members. Alternatively, the secondary beams may be attached to the stiffeners welded to the ends of the stubs (see Fig. 4.16(c)).

Generally, little advantage is gained from trying to achieve moment continuity between the stub girder and the adjacent columns. The main design criterion for stub girders is the longitudinal shear transfer between the chords via the stubs which is largely unaffected by continuity. Nevertheless, the bottom chord can be easily designed to develop a suitable connection to the columns (e.g. by end plates) and the slab reinforcement designed to resist the appropriate tension. This can be enhanced by the attachment of any T-section upper chord (as in Fig. 4.16(b)) to the column flange. The column web would usually have to be reinforced locally to resist the forces developed by these connections.

The requirements for local shear transfer at the stubs generally mean that it is necessary to introduce vertical stiffeners at the ends of the outer stubs (where shear forces are greatest). These stiffeners can often be omitted on the stubs towards mid-span.

4.3.2 *Design considerations*

The moments and forces in a stub girder can be (and have been) determined by various forms of 'elastic' numerical analysis. However, the approach adopted here is a simplified hand analysis at the ultimate and serviceability limit states. This approach ignores the bending resistance of the concrete slab, and is therefore conservative and leads to slightly greater forces in the chords than in reality. The applied load (including self-weight) is assumed to act through the secondary beams. The variation of moment and moment capacity along the beam is illustrated in Fig. 4.17.

Most of the resistance of a stub girder to global bending is achieved by composite action involving tension in the steel chord and

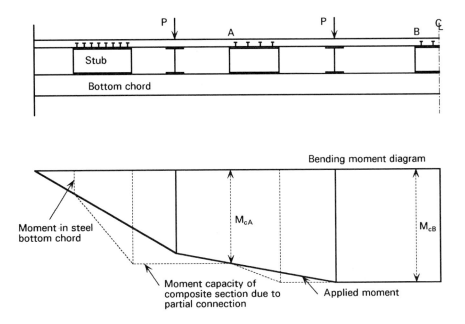

Figure 4.17   Build-up of moment capacity along stub girder.

compression in the concrete slab. In addition to this global action, the transfer of vertical shear forces across the web openings between the stubs causes local (or Vierendeel) bending in the chords. The combined moment capacities due to local and composite action should exceed the applied moment at all points along the span.

In the simplified hand analysis presented here, it will be assumed that all the Vierendeel bending is resisted by the bottom steel chord. This design approach is adopted as the member forces it predicts are in generally good agreement with the results of three full-scale tests. More shear connectors, and hence longer stubs, are normally required in the higher shear zones towards the outer parts of the span.

The moment capacity of the stub girder calculated using the simplified approach presented here should exceed the applied moment due to both dead and live loads (using the load factors in BS 5950: Part 1). Alternatively, a computer analysis may be used to determine the moments and forces in the top and bottom chords. The most commonly used is a plane frame approach, in which the various members are modelled by elements of calculated stiffnesses. This will result in a less conservative distribution of forces than that assumed above but does mean that the slab has to be designed for the forces that it attracts (see the section below on design of the concrete flange, page 147).

Moment capacity due to axial force in the bottom chord

The tensile resistance of the steel section is simply

$$R_s = p_y A \tag{4.1}$$

where A is the cross-sectional area of the steel bottom chord
        p_y is the design strength of steel (to BS 5950: Part 1).

The compressive resistance of the concrete slab is

$$R_c = 0.45 f_{cu} B_e D_{av} \tag{4.2}$$

where $0.45 f_{cu}$ represents the compressive strength of a stocky column (or wall) with concrete of cube strength f_{cu}
        B_e is the effective breadth of the slab considered to act with each beam (discussed below)
        D_{av} is the average depth of the slab in the case where the ribs of the slab run parallel to the stub girder, or the minimum depth in other cases.

The compressive resistance of a steel top chord may also be included in R_c.

The effective breadth of the slab is taken as one quarter of the span of the stub girder (or one eighth for edge beams) but not greater than 0.8 times the spacing between the stub girders. This limit of 0.8 is introduced in the design of primary beams because of the influence of combined slab and beam bending in the same direction in such cases. The same effective breadth is also recommended for use in serviceability calculations.

If $R_s < R_c$ then the plastic neutral axis of the composite section lies in the concrete. Conservatively, the lever arm is the distance from the mid-depth of the slab to the mid-depth of the steel bottom chord, D_{eff}. Hence, the moment capacity of the composite section is

$$M_c = R_s D_{eff} \tag{4.3}$$

If $R_s > R_c$, then the plastic neutral axis lies in the steel section, and usually in the top flange, such that

$$M_c = R_c D_c + R_s \frac{D}{2} \tag{4.4}$$

where D_c is the distance from the top of the steel bottom chord to
the mid-depth of the slab such that $D_c = D_{eff} - D/2$
D is the depth of the steel bottom chord.

The above formulae assume that 'full shear connection' is provided, so
that the full plastic moment capacity of the composite section can be
developed. In many cases, M_c will exceed the mid-span moment by a
considerable margin. This is necessary because the steel section should
also be able to resist local (Vierendeel) moments across the opening.

The moment capacity of the cross-section builds up in stages along
the stub girder resulting from the longitudinal force transferred via the
shear connectors at the stubs. It is also necessary to check the moment
capacity at intermediate locations, as shown in Fig. 4.17. These checks
are covered in the next section.

Longitudinal shear transfer

It follows that to achieve the moment capacity calculated in the previous section, the longitudinal force to be transferred between the
points of zero and maximum moment should exceed the smaller of R_s
or R_c.

The longitudinal shear transfer is normally achieved by provision of
shear connectors between the concrete and the 'stub', by shear in the
web of the stub, and finally, by welds or bolts between the stubs and
the steel chord. The number of shear connectors and bolts, and size of
welds is chosen to resist this force.

The rate of build-up of force in the concrete or steel broadly follows
that of the shear force diagram along the beam. Consequently, more
shear connectors, and hence longer stubs, are required in the high
shear zones towards the outer parts of the span. If the design capacity
of each of the shear connectors is P_d, it follows that the total number
of shear connectors needed in the *half*-span is

$$N = \frac{R_s}{P_d} \not> \frac{R_c}{P_d}$$

The characteristic resistance of the shear connectors may be
obtained from Table 5 of BS 5950: Part 3: Section 3.1. The design
capacities are obtained by multiplying these values by a factor of 0.8 in
the positive moment region.

The above approach applies for 'full shear connection' in the
composite section at the point of maximum moment. However, there
is scope for reducing the total number of shear connectors where the

moment capacity exceeds the applied moment. BS 5950: Part 3 permits the use of 'partial shear connection' for beams up to 16 m span.

If the total force transferred by the shear connectors from the adjacent support to the point on the span under consideration is R_q (such that $R_q < R_s$ and R_c), then

$$M_c = R_q D_c + R_s \frac{D}{2} \qquad (4.5)$$

where D_c is defined as used in Equation 4.4

This formula applies when the plastic neutral axis of the section lies in the top flange of the steel bottom chord, such that $R_q > R_w$, where R_w is the tensile resistance of the web.

The degree of shear connection, K, is defined as $K = R_q/R_s$ (when $R_s < R_c$) and R_q/R_c (when $R_c < R_s$). The minimum degree of shear connection at the point of *maximum* moment is given by

$$K \geq \frac{L-6}{10} \quad \text{and} \quad 1.0 \geq K \geq 0.4$$

where L is the beam span in metres.

Therefore, the minimum value of R_q needed in mid-span is determined, irrespective of the loading. The same approach may be adopted for the build-up of moment capacity along the beam as determined by the magnitude of R_q from the adjacent support to any point under consideration. Equation 4.5 applies when the web is fully in tension. A further equation may be derived when $R_q < R_w$, so that the web of the bottom chord is partly in tension, as follows

$$M_c = R_q D_{eff} + M_s - \frac{R_q^2}{R_w} \frac{D}{4} \qquad (4.6)$$

where M_s is the moment capacity of the steel section.

This is an exact equation based on 'stress block' analysis and represents the *maximum* available moment capacity of the section. The first term represents the moment due to tension in the bottom chord (as given by R_q), and the second term is that due to pure bending in the chord. The final term represents the adverse effect of tension on the moment capacity of the steel section.

Equation 4.6 applies where R_q is relatively small, i.e. close to the supports. In the limit, M_c tends to M_s and R_q is zero. The effect of high vertical shear should also be considered (see next section) by reducing the term R_w.

The distribution of shear connectors should be such as to ensure that the value of M_c (determined from Equations 4.5 and 4.6) exceeds the global applied moment at all points. Critical cross-sections are at the higher moment side of the openings, shown as points A and B in Fig. 4.17. The moment capacity remains constant between the stubs in the absence of any other means of shear transfer.

As a 'safe' simplification, the total number of shear connectors needed at any stub may be distributed in proportion to the shear force diagram along the beam. This determines the number of shear connectors positioned over each stub relative to the total number required in the half-span. A nominal number of shear connectors (only 1 every 450 mm) is appropriate in zero-shear zones.

The shear connectors may be arranged singly or in pairs along the stubs subject to minimum spacing criteria of 4ϕ laterally and 5ϕ longitudinally (where ϕ is the stud diameter). These requirements effectively determine the minimum length of stub to be used. Additional transverse reinforcement in the slab is needed to enhance a smooth transfer of shear into the slab (see the section below on transverse reinforcement in slab).

Design of the steel bottom chord

The internal forces developed in a stub girder are presented in Fig. 4.18. Longitudinal forces are transferred discreetly at the stubs rather than gradually along the beam. Equilibrium is satisfied by

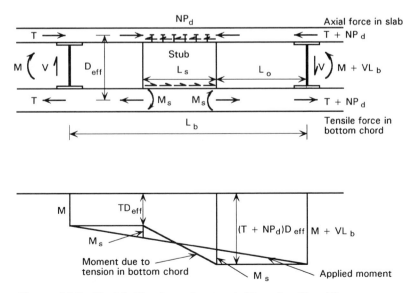

Figure 4.18   Model of load transfer at stub (Type A or Type B).

development of moments in the bottom chord. Therefore, the bottom chord is subject to tension arising from the global (or primary) bending action and also to local (or secondary) moment and shear arising from the shear forces applied to the girder. The relative magnitude of these effects depends on the ratio of moment and shear at any point along the span (see Fig. 4.18).

The steel chord should be a 'plastic' or 'compact' section in accordance with BS 5950: Part 1. This is to ensure that it can develop its plastic moment capacity. A significant amount of rotation capacity is not required.

Vierendeel moments

The shear conditions around an opening demand the greatest consideration. This behaviour is similar to that occurring when a large opening is positioned in the web of a girder such that the upper flange and upper part of the web are cut away. The transfer of vertical shear force across the opening is then resisted by local bending of the bottom chord and the slab. This is commonly known as 'Vierendeel bending'.

The high relative stiffness of the bottom chord to the top chord (the concrete slab) means that most of the Vierendeel bending moment is resisted by the bottom chord and the contribution of the slab can be ignored. For equilibrium, the moment difference between the edges of adjacent stubs is dependent on the shear force transferred. Therefore, the Vierendeel moment in the bottom chord, M_v, adjacent to the stubs is given by

$$M_v = VL_o \qquad (4.7)$$

where L_o is the distance between the edge of the stub and the point of contraflexure in the bottom chord
V is the applied shear force across the opening.

As a first approximation, L_o may be assumed to be mid-way between the edges of the adjacent stubs. Hence, in Figs 4.16(a) and (b), L_o is the length of the opening between the stub and the secondary beam. In Fig. 4.16(c), L_o is half the distance between the adjacent stubs. At the outer openings, L_o is the distance from the outer stub to the adjacent column.

The local vertical force applied to the bottom chord by the secondary beam also causes local moments in the bottom chord. However, these additional effects are accounted for if V is defined as being the maximum shear force across each opening, and each opening is checked separately.

The bottom chord should be able to maintain equilibrium by ensuring that the moment capacity of the steel section, M_s, exceeds M_v adjacent to each stub. The moment capacity of the steel section, M_s, should also take into account the influence of shear and tension, as considered in the following sections.

Influence of shear

The shear force, V, is considered to be resisted entirely by the web of the bottom chord, because the slender slab between adjacent stubs may not be able to resist significant shear force. Alternatively, a plane frame computer analysis may be used to determine the moments and forces directly in the top and bottom chords. This will result in a less conservative distribution of forces than that assumed, but does mean that the slab is to be designed for the forces that it attracts (see the section below on the design of the concrete flange, page 147).

The shear stress applied to the web of the bottom chord reduces the effectiveness of the bottom chord in bending and tension. This may be taken into account by modifying the effective thickness, t_e, of the web according to

$$t_e = t_w \sqrt{\left[1 - \left(\frac{V}{V_u} \right)^2 \right]} \tag{4.8}$$

where V_u is the ultimate shear strength of the web which is equivalent to a shear stress $0.6 p_y$ applied uniformly over the full depth of the section (as in BS 5950: Part 1 clause 4.2.6)

t_w is the actual thickness of the web.

In principle, this formula is rather less conservative for sections subject to high shear than the linear interaction formula presented in BS 5950: Part 1. It is presented as an alternative for highly stressed sections. However, it is more conservative for sections subject to low shear, and hence no reduction in web thickness need be taken when $V < 0.6V_u$ as in BS 5950: Part 1 (see Fig. 4.19).

The effective web thickness, t_e, is now used to recalculate the properties of the steel bottom chord, i.e. R_w, R_s and M_s, as defined previously in the section on longitudinal shear transfer. These properties also influence the moment capacity of the composite section, M_c.

Influence of tension

The influence of tension is already included when calculating M_c using Equations 4.3–4.6. These are 'exact' equations and represent the maximum capacity of the composite section. Overall equilibrium

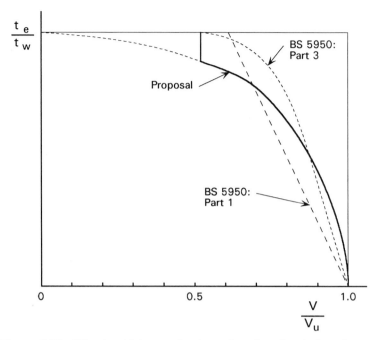

Figure 4.19   Effective thickness of web as a function of web shear force.

should be satisfied by ensuring that the moment capacity at all points exceeds the applied moment (see Fig. 4.17).

However, the moment capacity of the steel section, M_s, is influenced by axial tension, T, which in turn influences the resistance to Vierendeel bending. This is best illustrated by considering the combination of stress blocks in the bottom chord at the low and high moment sides of the opening, as shown in Fig. 4.20.

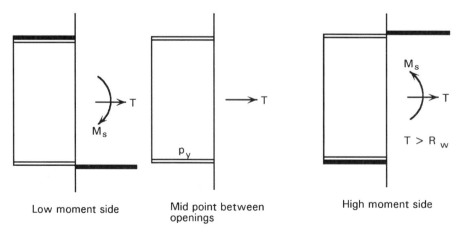

Figure 4.20   Combination of stress blocks in bottom chord between stubs.

The proportions of the section not utilised in resisting tension can be used to resist the Vierendeel bending effect (shown shaded in Fig. 4.20). The interaction between moment and tension is then of the form illustrated in Fig. 4.21. Where the web area is small, a linear interaction is appropriate. This is equivalent to a direct combination of bending and axial stresses. The linear interaction approach becomes more conservative (by about 10%) as the web area increases. However, the reduced moment capacity of the steel section, $M_{s,red}$, may be determined with reasonable accuracy from

$$\frac{M_{s,red}}{M_{s,eff}} = 1 - \frac{T}{p_y A_{eff}}$$

(4.9)

where T is the applied tensile force at a given location. This is equivalent to the force, R_q, transferred via the shear connectors from the support to the point under consideration.
A_{eff} is the effective area of the section (including t_e) $\leq A$
$M_{s,eff}$ is the moment capacity of the section (including t_e) $\leq M_s$
t_e is the effective thickness of the web (calculated using Equation 4.8).

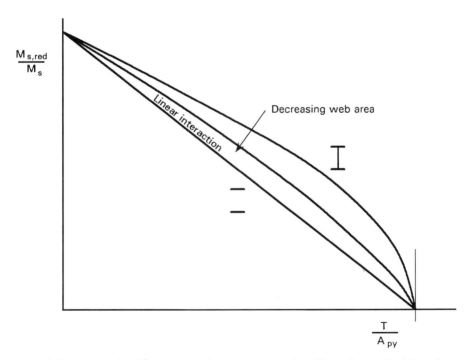

Figure 4.21   Interaction between plastic moment capacity of I-section and axial tension.

Therefore, for satisfactory design of the bottom chord, $M_{s,red} > M_v$ (see Equation 4.7) and $T < p_y A_{eff}$. As a first approximation, the web area can be ignored in calculating the effective section properties.

Local design of the stub

Equilibrium of forces on the stub dictates that there must be a vertical reaction developed between the base of the stub and the bottom chord. This local behaviour is illustrated in Fig. 4.22. The magnitude of this separation force per unit length is

$$f = N_i P_d \left(\frac{6D_s}{L_s^2} \right) \tag{4.10}$$

where N_i is the number of shear connectors, of design strength P_d, attached to a stub section of length L_s and depth D_s.

The welds are therefore designed to resist longitudinal shear and uplift. These compression and shear forces may be combined vectorially to give a maximum force per unit length of weld of

$$f_{max} = \frac{N_i P_d}{L_s} \left[1 + \left(\frac{6D_s}{L_s} \right)^2 \right]^{0.5} \tag{4.11}$$

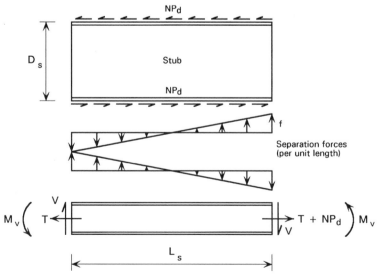

Figure 4.22   Local forces on stub and bottom chord.

This force is used in the design of the fillet welds attaching the stub to the lower chord. Similarly, connecting bolts, which can be used as alternatives to welds, must be capable of resisting these combined forces. Friction-grip bolts will usually be needed to avoid the effects of slip on deflections.

The web of the stub is designed to resist the compression force, f_{max}, and a longitudinal shear force as transferred via the shear connectors. The web slenderness is $2.5D_s/t$ when subject to buckling, and its compressive strength can be evaluated as for a strut, according to Table 27 of BS 5950: Part 1. Additionally, shear and compression stresses are combined vectorially using the von Mises criterion and should be less than p_y.

In many cases it is necessary to stiffen the edge of the stub using a vertical welded stiffener. This load-bearing stiffener should be designed to resist a force equivalent to $N_i P_d D_s/L_s$ (i.e. ignoring the contribution of the web). It is not economic to stiffen the web of the bottom chord and therefore it should be checked for its resistance to web bearing or buckling when subject to this force.

Design of the concrete flange

The concrete slab acts as the compression flange of the stub girder. It behaves effectively as a strut (or more correctly a braced wall) which is restrained at the attachments to the stubs and secondary beams. In theory, the flange behaves as a 'stocky' column or strut provided the ratio of its unsupported length to slab depth does not exceed 12. The real behaviour is rather different in that the slab is not continuously restrained across its width, and also there is some small flexibility of the attachments of the slab to the shear connectors (see Fig. 4.23). Local moments and shear forces may also develop in the slab due to its stiffness, but these are usually ignored. Local uplift forces on the shear connectors may also be ignored, provided the deformation across the opening is small (see the section below on deflections, page 150).

Figure 4.23   Stability of compression flange between stubs.

In the absence of other detailed guidance, it is considered necessary to restrict the maximum unsupported length of slab between longitudinal restraints to a span-to-depth ratio of 10, based on the average slab depth, D_{av}. Assuming that the average depth is 100 mm, the maximum unsupported length of slab when subject to its design compressive stress is therefore 1000 mm. This is recommended as a reasonable rule of thumb to be used when sizing stub girders. Secondary beams also act as effective restraints to the slab. This means that the maximum distance between the edges of adjacent stubs in Figure 4.16(a) and (b) is 2000 mm.

This method may be extended to treat slabs with lower axial forces by multiplying the span-to-depth limit of 10 by R_c/R_s for $R_c > R_s$ (implying that the slab is not fully stressed to $0.45f_{cu}$). However, the coincident influence of shear tends to cause eccentricity of load, and hence exacerbates the instability of the concrete flange. It is suggested that the slab span-to-depth ratio of 10 is retained for high shear regions irrespective of the force in the slab. The presence of an upper T-chord would also have a stabilising effect and may be used to increase the unsupported length of the slab between restraints.

Where the analysis determines the moments and forces in the slab (e.g. by plane frame analysis), the slab should be designed to resist these forces. Moment and axial forces may then be combined using the column design charts of BS 8110.[14] This often necessitates the provision of additional reinforcement in the slab and more shear connectors at the edges of the stubs to resist local uplift forces.

Transverse reinforcement in slab

In order to develop a smooth transfer of force from the shear connectors into the concrete, it is necessary to provide adequate transverse reinforcement (i.e. transverse to the axis of the beam). This can be achieved by straight bars or mesh, but more efficient detailing arrangements using 'herring-bone' reinforcement have been developed (see Fig. 4.24).

The resistance to longitudinal shear may be evaluated by considering the potential shear planes on either side of the line of shear connectors. The resistance per unit length is defined in BS 5950: Part 3: Section 3.1, clause 5.6.3 as

$$V = 0.7A_{sv}f_y + 0.33\eta A_{cv}f_{cu} < 0.8\eta A_{cv}\sqrt{f_{cu}} \qquad (4.12)$$

where A_{sv} is the cross-sectional area of reinforcement per unit length of the beam for each shear plane

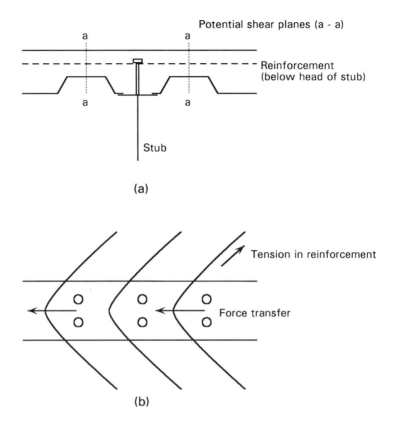

Figure 4.24 Influence of transverse reinforcement in controlling longitudinal shear failure. (a) Illustration of longitudinal shear failure. (b) Use of
'herring-bone' reinforcement in slabs.

A_{cv} is the mean cross-sectional area of concrete per unit length

$$\eta = \begin{cases} 1.0 \text{ for normal weight concrete} \\ 0.8 \text{ for lightweight concrete} \end{cases}$$

f_y is the design strength of the reinforcement.

For internal beams, the resistance V may be *doubled* to account for
formation of two shear planes. The contribution of the decking can be
included, as suggested in BS 5950: Part 3: Section 3.1, if the decking
runs perpendicular to the stub girder, or the longitudinal edges of the
decking are fastened together (e.g. by screws or stitch welding).
Generally, the effect of the decking is not included in stub girder
design. The upper limit on V is introduced to prevent local crushing of
the concrete.

The applied force per unit length is conservatively given by $N_i P_d/L_s$
(see the previous section on local design of the stub, page 146). It
follows that this force should not exceed the available longitudinal

shear resistance provided over the stubs, as determined by Equation 4.12. It is usually found that significant amounts of additional bar reinforcement are needed in the region of the stubs to control this effect. However, the full scale tests showed that the zone of longitudinal shear transfer is no longer than the length of the stub and that Equation 4.12 is very conservative.

Construction condition

In the construction stage, the steel bottom chord is designed to support the self-weight of the floor and other construction loads (taken as equivalent to $0.5\,\text{kN/m}^2$ in the design of the beams). The chord may be checked in bending in accordance with BS 5950: Part 1 (using the load factors in Parts 1 and 3). It is usually found that one or two vertical props are needed so that the moments in the chord are reduced.

In-built stresses during construction do not affect the final collapse of the stub girder, which is assessed on the basis of factored dead and imposed loads applied to the composite section (see the previous section on moment capacity due to axial force in the bottom chord, page 138). This implies that significant redistribution of internal stresses occurs. Serviceability stresses are not normally calculated in this form of construction because local yielding does not have a major effect on deflections.

When the system shown in Fig. 4.16(b) is employed, the possibility exists of designing the steel top chord for the compressive forces induced due to the bending moment developed in the girder at the construction stage. The top chord is then designed as a strut acting over the lengths of the openings. This system can avoid the need for temporary propping.

Deflections

The deflection of a stub girder system comprises a component caused by normal bending action and a component caused by the transfer of shear across the openings (Vierendeel action). Deflections are normally calculated for unfactored imposed load and should be limited either by propping, by introducing a top chord, or by pre-cambering.

The bending deflection is obtained by calculating the effective second moment of area of the section. The area of concrete in the composite section is reduced to an equivalent area of steel by dividing by the modular ratio of steel to concrete, α_e. The appropriate values of α_e are given in BS 5950: Part 3, but 'average' values of 10 for normal weight concrete and 15 for lightweight concrete are generally used for buildings with low permanent imposed loads (i.e. normal stage).

The second moment of area of the combined top and bottom chords, ignoring the contribution of the stubs, is

$$I_c = \frac{AA_c/\alpha_e}{(A + A_c/\alpha_e)} \times D_{eff}^2 + I_s \tag{4.13}$$

where $A_c = D_{av}B_e$ plus the contribution of any steel section embedded in the concrete

D_{eff} is defined as the distance from the mid-depth of the slab to the mid-depth of the steel bottom chord

I_s is the second moment of area of the steel bottom chord of cross-sectional area A.

Hence, the mid-span bending deflection due to uniform imposed load is obtained directly from

$$\delta_b = \frac{5W_iL^3}{384EI_c} \tag{4.14}$$

where W_i is the total unfactored imposed load on the beam of span L

E is the elastic modulus of steel ($= 205$ kN/mm^2).

The mid-span 'shear' deflection can be estimated by considering the deflection due to local Vierendeel bending action in the bottom chord such that

$$\delta_v = \Sigma \frac{VL_o^3}{3EI_s} \approx \frac{W_iNL_o^3}{24EI_s} \tag{4.15}$$

where V is the shear force per opening

L_o is the length of the opening defined in the previous section on the design of the steel bottom chord or Fig. 4.18. Correctly, L_o is the distance from the stub to the point of contraflexure in the chord.

The summation is over all the openings in half the beam span. For a regular distribution of openings, the deflection tends to the second formula (where N is the number of openings in the span). In order to avoid excessive distortion of the opening and uplift forces on the shear connectors, it is proposed that $\delta_v/L_o < 10^{-3}$.

The total central deflection is therefore $\delta_b + \delta_v$. It is usually found that the shear term is much less ($< 10\%$) than the bending term. This calculation is conservative because the bending stiffness of the slab has been ignored. If the stub girder is propped, then the self-weight deflection on removal of the props can be calculated in a similar manner (using the long-term modular ratio as in BS 5950: Part 3).

Secondary beams

Secondary beams directly support the composite slab and can be designed either as simply supported or continuous composite beams (see Fig. 4.17). Continuous design is appropriate where the secondary beams pass over the bottom chord. However, special consideration is required along the column lines where it may not be possible to achieve continuity. In these cases it may be necessary to use a heavier beam.

For simply supported beams, the design tables in Reference 15 may be used for rapid selection of the beam size. For continuous beams, elastic global analysis or plastic hinge analysis may be used to determine the moments and forces in the section. This approach is presented in BS 5950: Part 3.

Plastic hinge analysis is commonly used for 'plastic' steel sections, and satisfactory design of beams with one degree of continuity (i.e. end bays) is obtained when

$$M_p + 0.5 M_n \left(1 - \frac{M_n}{8 M_u} \right) \geq M_u \tag{4.16}$$

where M_n is the negative (hogging) moment capacity of the composite section (including reinforcement, but ignoring mesh)
M_p is the positive (sagging) moment capacity of the composite section
M_u is the free bending moment at the ultimate limit state.

This method of analysis may result in considerable redistribution of moment at the serviceability limit state due to the early formation of plastic hinges at the supports. The effect is to cause increased deflections under repeated loading. To avoid these so-called 'shakedown' deflections, the ratio of M_p/M_n should not exceed 1.5. This often dictates the sizing of the steel secondary beams. Typically, when M_p/M_n equals 2.0, the additional deflection developed as a result of shakedown effects is approximately 30%. Elastic deflections should also take into account the influence of pattern loads.

As a rule of thumb, adequate serviceability performance is achieved when the span-to-depth ratio of simply-supported beams is less than 20, and that of continuous beams (end bays) is less than 24. The structural depth in this case is the combined slab and steel beam depth. Detailed analysis of continuous composite sections is covered in BS 5950: Part 3.

Vibration effects

The design procedures for vibration are covered in Chapters 2 and 3. Further design information can be obtained from Reference 16.

4.3.3 Design procedure for stub girders

The design procedure is set out in three stages representing broadly the sequence used in design. It is assumed that the proportions of the stub girders are typical of good practice. Design should conform to the requirements of BS 5950: Parts 1 and 3, where relevant.

Construction condition

(1) Determine the moment diagram under factored loads during construction, allowing $0.5\,kN/m^2$ for imposed construction loads.

(2) Select the size of the bottom chord. Determine its moment capacity, M_s. Compare with the applied moment in (1). If not adequate (as is generally the case), use temporary props to reduce the moment, or alternatively use a steel top chord. If props are used, compare the moment capacity with the reduced moment.

(3) If a steel top chord is used, determine the compression in the chord when subject to the maximum moment used in (1). Check its adequacy as a strut. Further checks on the bottom chord are included in the next section on ultimate loads.

(4) Calculate the deflection of the stub girder after construction (ignoring construction loads) based on the steel section (including the top chord if used).

Ultimate loads

(1) Determine the moment and shear force diagrams at ultimate loads (factored self-weight, dead and imposed loads).

(2) For the selected bottom chord, calculate its moment and shear capacity (see construction stage). Check that the shear capacity exceeds the maximum shear force, V. As a first approximation, the maximum length of the opening, L_o, is such that $M_s > VL_o$ (see the section above on the design of the steel bottom chord,

page 141) and $L_o/D_{av} \leq 10$ (see the section above of the design of the concrete flange, page 147), where D_{av} is the average slab depth.

(3)   Calculate the moment capacity of the composite stub girder section assuming full shear connection (see Equation 4.3 or 4.4). Check that the moment capacity of the stub girder exceeds the maximum moment determined in (1).

(4)   Calculate the number of shear connectors needed in the half span for full shear connection (see the section above on longitudinal shear transfer, page 139). Distribute in accordance with shear force diagram. Calculate the number of shear connectors that can be placed along the stubs at the minimum recommended spacing. If this is not greater than the required number, consider partial shear connection design (see (5) below). Determine the minimum degree of shear connection for the span of the girder.

(5)   For the required degree of shear connection, determine the moment capacity of the stub girder (see Equation 4.5 or 4.6). Check that the moment capacity exceeds the maximum moment determined in (1). If adequate, then proceed. If not, increase the depth of girder or the size of the bottom chord.

(6)   Check the adequacy of the stub girder at *all points* along the span. This is based on a sensible distribution of shear connectors as in Step (4). Use Equation 4.5 or 4.6 to determine the moment capacity, inserting the appropriate value of R_q at the point under consideration. Critical cross-sections are at the higher moment side of the openings. Ensure that the moment capacity exceeds the applied moment at all points in the span.

(7)   For the selected opening widths, determine the Vierendeel bending moment (see (2) above). Modify the moment capacity of the steel section according to the tension and shear forces existing at the low moment side of the openings (see Equations 4.8 and 4.9). Re-check the ability of the steel section to resist Vierendeel bending. If not adequate, reduce opening width. Check all openings.

(8)   Check the local design of the stub (see Equation 4.11). Use vertical stiffeners for the outer stubs. Increase the length of the stub to reduce the web stresses or to attach more shear connectors.

(9)   Check the adequacy of transverse reinforcement in the slab (see Equation 4.12).

(10) Determine the weld size or number of bolts between the stubs and the bottom chord.

Serviceability

(1) For the proportions of semi-permanent and permanent imposed loads, determine the appropriate modular ratio to be used in deflection calculations. Note that no stress checks are made at serviceability as it is assumed that any local yielding has little effect on deflections.

(2) Calculate the second moment of area of the composite section (see Equation 4.13). Calculate the central bending deflection under imposed load.

(3) Calculate the 'shear' deflection due to Vierendeel bending effects (from Equation 4.14). Sum the bending and shear deflections. Check that the shear deflection across an opening does not exceed $L_o/1000$. Check that the total imposed load deflection does not exceed span/360 or 30 mm (sensible maximum).

(4) Add self-weight and imposed load deflections (for self-weight deflection, see construction condition step (4)). Limit the total deflection to span/200 or 70 mm (sensible maximum). Consider pre-cambering the bottom chord.

(5) Determine the natural frequency of the stub girder (see the previous section on vibration effects, page 153). If this exceeds 4 cycles/second, no further calculations are required. If not, but the natural frequency exceeds a minimum of 3 cycles/second, further calculations are required to establish the response of the floor.[16]

4.3.4 Conclusions

Stub girders are appropriate for rectangular grillage systems with spans of 12–20 m and secondary beam spans of 8–12 m. Various girder configurations are possible and the use of an upper chord can in some cases overcome the need for temporary propping.

Failure often occurs local to the outer stubs, such as by compression or shear failure of the stubs, or longitudinal shear failure of the slab adjacent to the stubs. This emphasises the need to consider local stiffening of the stubs and additional transverse reinforcement in the slab around the shear connectors. In other cases, combined bending and tension on the bottom chord controls the design.

Although stub girders have been used in the UK, the author considers that they have a limited use. However, where large spans are required with relative restricted construction depths they are a useful alternative to consider.

Stub girders are also design intensive with a high degree of fabrication required. In today's market, this last remark is of importance when comparing this method of construction with other structural forms.

4.4 The parallel beam approach

The aims of the parallel beam approach are:

- to reduce fabrication and erection complexities by reducing the total number of members in a steel frame
- to reduce the weight of the steel beams by use of continuity
- to reduce the complexities and costs that occur at intersections between structural members and between structural members and services.

The successful achievement of these aims produces a steel frame that can be constructed quickly, easily and cheaply. Most importantly, the resulting building has great flexibility of services and planning, allowing speedy erection of the frame and reducing the overall building cost.

A typical general arrangement of this method of framing is shown in Fig. 4.25. To avoid clashes between services and/or structure, it has two parallel planning zones, one above the other. The services are then arranged 'parallel' to the structural depth.

4.4.1 *Conventional approach*

In a three-dimensional orthogonally framed structure, conventionally, beams in the X and Y directions (i.e. along horizontal axes) are in the

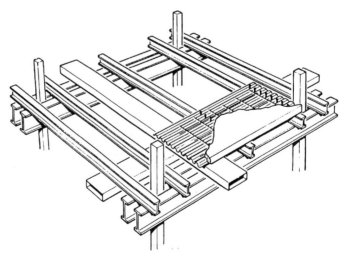

Figure 4.25 Parallel grillage system showing service zones.

same XY plane (i.e. at the same level) and are supported by columns in the Z direction. The beams in the X or Y direction are generally located at centres which are a function of the Y and X dimensions of the column grid. The location of beams is therefore dictated to a large degree by the column layout.

4.4.2   Parallel beam approach

The parallel beam approach to framing also requires members in the X, Y and Z directions, but the beams in the X and Y direction are displaced in the Z direction relative to each other (i.e. not on the same level). With a beam in one direction passing over the beam in the other direction, beam intersections in the same plane are therefore avoided, which greatly simplifies the connections between the members (Fig. 4.26).

Location of spine beams

The spine beams are displaced laterally so that they pass beside the columns, thus avoiding an intersection with the column. They do not connect to the floor slab and are therefore designed non-compositely.

Internally, twin parallel beams are normally used, whereas, externally, a single beam only is used. Spine beams are displaced either side of the column with 20–40 mm gaps between the face of the column and edge of the spine flange. The spine beams are connected via brackets to the columns.

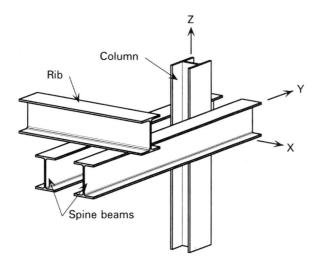

Figure 4.26   Displacement of rib, spine beams and column to simplify connections.

UNIVERSITY GLAMORGAN
PRIFYSGOL MORGANNWG
Learning Resources Centre

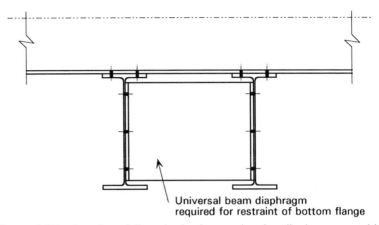

Universal beam diaphragm
required for restraint of bottom flange

Figure 4.27 Junction of rib and spine beams showing diaphragm to stabilise the latter.

Ideally, the distance between the inner faces of the webs of the twin
spines should be equal to the overall depth of a standard UB section
such as 533 × 210 UB. This allows diaphragms to be cut from the UB
section and bolted between the spines directly under the ribs located
with the side of the column as shown in Fig. 4.27. Diaphragms may
also be necessary between ribs at mid-span to stabilise the top flange
under sagging moments.

The spine beams are therefore braced together at the column and at
an appropriate distance either side of the column to provide lateral
and torsional restraint to the hogging moment region of the spine. The
top flanges of the spines are braced together and held in position by
bolting to the bottom flange of the ribs. The structural efficiency of the
spine beams is improved by making them continuous.

Location of ribs

Generally, it is preferable for the ribs to be of the same section, even
when spans and/or loading vary. This rationalisation simplifies vertical
setting out and detailing. It also minimises price/tonne and delivery
times.

While it is necessary for the spine beams to be located adjacent to
the columns, the beams in the other direction (ribs) are displaced
laterally to miss the column, as shown in Figs 4.25 and 4.26. Their
lateral disposition is determined by economics or the planning
requirements of the floor and not necessarily the column grid. Their
spacing should be such that propping of the deck is not required.

By avoiding intersections, beams can now be more than one span in
length without recourse to complex jointing, dependent only on the

length that can be obtained and economically handled. Thus beams no longer need notching or end plating.

These rib beams are connected to the floor slab and are designed compositely. The structural efficiency is also enhanced by developing continuity.

4.4.3 Detailed aspects of parallel beam framing

Beams

As a general rule, it is better to let the composite and more lightly loaded rib beams span the greater distance, with the more heavily loaded spine beams spanning the shorter distance. However, this arrangement may need to be reversed for very long span ribs if propping is to be avoided or to suit layout of services (see Section 4.4.10).

Column orientation

The major axes (XX) of the columns are normally aligned parallel to the spine beams, as shown in Fig. 4.26. Thus any beam rotation at the supports, due to differential loading on adjacent spans, induces weak axis bending in the column. Column moments and shear forces on the supporting brackets are thereby minimised.

Where single spine beams are used around the perimeter of the building, as shown in Fig. 4.28, the same orientation ensures that the column moment from the eccentric beam support is applied

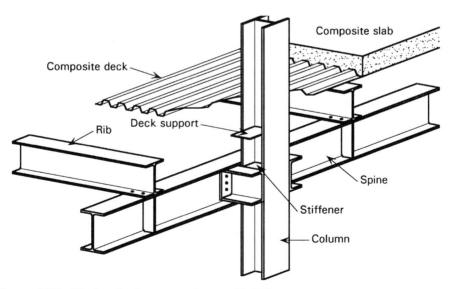

Figure 4.28   Single spine beam at perimeter of building.

about the former's major axis. This orientation also produces the most favourable slenderness ratio for the column. However, in certain circumstances, for example, where services are required to pass down between the column web and the spine beam, it may be necessary to orientate the columns so that their minor axes are parallel to the spine beams. It should be appreciated, however, that this orientation will attract larger column moments, possibly requiring stronger support brackets. There will also be a reduction in the degree of the column repetition.

Connections

Connections between the ribs and the spine beams are made by bolting the bottom flange of the upper beam to the top flange of the lower as shown in Figs 4.26 and 4.28. Connections between beam and column are made by bolting through the beam web to brackets welded to the columns, as shown for a single spine beam in Fig. 4.28. (For a double spine beam the brackets are symmetric about the column major axis.)

Differential deflections

The distance between an external edge beam and the first internal rib will be half the normal rib spacing because the rib beams are displaced to miss the columns as indicated in Figs 4.25 and 4.26. This rib (assuming all ribs are of similar section) will not have the same magnitude of deflection as an internal rib. The transition from no deflection adjacent to an external column, to full deflection at the middle of a long-span rib beam will be eased thereby.

In some situations, it may be necessary to further ease the transition by the introduction of semi-flexible structural members between the external short-span edge beams and the internal long-span ribs. This increases the length over which the change in deflection occurs, thus decreasing the rate of change of slope.

These semi-flexible members are often in the form of 'goal posts', as shown in Fig. 4.29. They allow use of the space between beams for services, while providing lateral and torsional restraint to the beams during erection. They will also provide stability during floor construction when the concrete has been poured but has not hardened.

Length of beams

The repetitive use of members of the same size, the need to achieve structural continuity and the constraints of the column grid determine the length of the rib and spine beams. For most efficient use, the

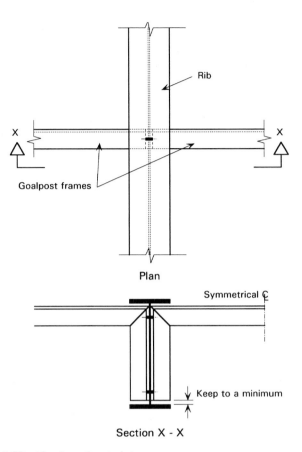

Figure 4.29 'Goal post' restraints.

composite members are made continuous over two spans, but occasionally three spans may be necessary when odd numbers of bays occur. In three-span rib beams, the centre span is often subject to hogging over its length under pattern loads with the central span unloaded. Structural efficiency of three-span beams can be enhanced by making the internal span longer than the outer spans.

In determining the length of beam to be used, consideration must also be given to the transportation, fabrication and erection processes. The beam may require turning on its side, lifting and handling in restricted spaces. There are special restrictions on transportation of beams over 27.4 m long. Plastic sections are anyway a requirement of the parallel beam approach, as currently conceived; these have the advantage of being easier to handle. As a guide, the minimum flange width should be approximately 1/125 of the length of the beam to be handled.

From experience, the maximum span for a three-span, composite rib is approximately 6 m, i.e. a total length of 18 m, and for a three-span spine beam is approximately 8 m, a total of 24 m. For two-span spines and ribs, maxima are about 12 m spans. Longer span beams may need jointing. This is normally achieved by full strength butt welding on site, the weld being located over a spine beam. Site welding is discussed in Section 4.4.8. As an alternative, bolted splices at positions of contra-flexure can be used to achieve continuity in the design of the beam.

4.4.4 *Construction aspects*

The time taken to hoist a 15 m long beam is normally the same as for a 7.5 m beam. On large sites 'hook time' may constitute the critical path for the construction programme. The parallel beam arrangement can lead to savings in construction time.

As a way of quantifying erection efficiency, an erection factor has been introduced, defined as:

$$\text{erection factor} = \frac{\text{number of pieces of steel} \times 100}{(\text{floor area in metres})^2}$$

Figure 4.30(a) shows the traditional layout of part of a floor to a multi-storey building having a 7.5 m square grid. The positions of the primary and secondary beams are pre-determined by the column grid.

Figure 4.30(b) shows the alternative parallel beam layout which has ribs at wider spacing but half the number of beams, albeit twice as long. Additionally, accumulative tolerance problems in multiple lengths of beams are reduced.

4.4.5 *Fabrication*

End plates and beam notches are no longer required; beam lengths now relate to grid centre lines, as opposed to the distance between the faces of column or beams. Beam lengths are therefore the same even though the supporting column or beam section size may have changed. Fabrication is simplified and repetition of members increased. The external and internal spines may often have the same section, this increases repetition.

4.4.6 *Erection*

With a large reduction in the number of lifts and increased repetition, combined with a system in which beams are 'landed on' rather than suspended between supports, erection is much simpler and faster. Alignment of bolt holes can be achieved more safely.

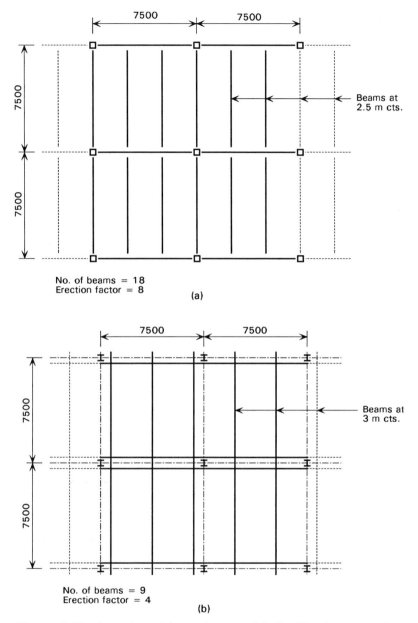

Figure 4.30 Examples of beam layouts. (a) Traditional construction. (b) Parallel beam approach.

4.4.7 *Services*

In conventional framing, a service zone is provided under the steelwork. However, in the parallel beam approach, services can be dealt with in a similar manner to the beams by being split into two orthogonal zones. Thus, in one direction, the services are located parallel to and in the same zone as the rib beams, and in the other, they are parallel to and in the same zone as the spines. By this arrangement the whole of the structural depth is available for service runs and clashes between services and/or structure can be avoided. This is illustrated in Fig. 4.25. Considering the overall depth of the floor zone, buildings framed in this way generally have the same or less depth overall than conventional framed construction.

4.4.8 *Site welding*

Where long-span continuous beams are required, it may be necessary to consider the need for site connections. This can be done conveniently by site welding. Provided the welding is properly specified and tested, experience has shown that site welding can be carried out to the same standard as shop welding and frequently the standard is higher. Precise details of the welding procedures and preparation should be agreed by the engineer with the main contractor, steelwork fabricator, welding sub-contractor and the Testing Authority prior to commencement of fabrication. Test pieces of the actual sections and preparation should be carried out and tested for compliance with the specification. Any necessary revisions to the procedure and preparation may then be incorporated and retested.

The actual cot is obviously dependent on the number of welds carried out, access, etc. To date, site welds have been successfully carried out on sections up to 914×305 UB.

4.4.9 *Lateral stability of frame and beams*

Lateral stability of the frame in both the horizontal (X and Y) directions is normally provided by the floor beams and slab transmitting the horizontal shear forces to vertical bracing or stiff cores or shear walls. Alternatively, the spine beams and columns may be considered as multi-bay portals to resist the horizontal load in the Y direction. In this case, vertical bracing is required for the plane parallel to the rib beams, i.e. in the X direction.

Ideally, the structure providing the permanent lateral stability should also provide the temporary stability during construction. In composite construction, the in situ floor slab is not able to transmit in-plane forces until the concrete has reached the required strength.

In the temporary conditions, it is important to ensure the stability of the structure under the weight of wet concrete and other construction loads, and also wind forces. The steel decking should be properly fixed to the beams in order for it to act as a shear diaphragm. Where shear forces in the diaphragm are high, seam fasteners should also be used (these have the added benefit of preventing differential displacement between sheets during concreting). Where two edges of a sheet occur over a beam, the studs should be through the deck, welded in a staggered pattern to ensure adequate connection of both edges.

Pattern loading of wet concrete should be considered to determine the length of the adjacent beam subject to negative bending. Restraint can be provided to the top and bottom flanges of ribs by the use of goal posts as shown in Fig. 4.29. Temporary vertical bracing may also be required until the concrete slab has sufficient strength to span between the positions of permanent vertical restraint.

4.4.10 *Propping during construction*

The composite rib beams should be checked for strength and deflection in the non-composite condition when subject to the wet weight of concrete and the construction loading (taken as $0.5\,kN/m^2$). From experience, it has been found necessary to prop composite beams supporting 130 mm thick lightweight concrete slabs for spans in excess of 9 m. There is of course the option of providing heavier and/or deeper sections if it is important to avoid propping during construction.

If propping is inconvenient to the construction process, the framing may be arranged so that the composite rib beams span less than 9 m, with spine beams having larger spans. This arrangement will save propping and rib costs at the expense of additional spine steel. It should be remembered that the deflections of slab, ribs and spine beams are accumulative.

4.5   Cellular beams

4.5.1 *Introduction*

This section will concentrate on the construction techniques for cellular beams which incorporates a series of circular openings in the web. The use of composite cellular beams for long-span construction is now well established and recognised for its suitability at keeping floor zones to a minimum by passing the services through the circular openings. The structural method of analysis for this form of construction is relatively complex and is not amenable to simple hand calculations. In view of this, the SCI was commissioned to develop a computer program covering cellular beam design. The software is available from Westok Structural Services Limited. Further

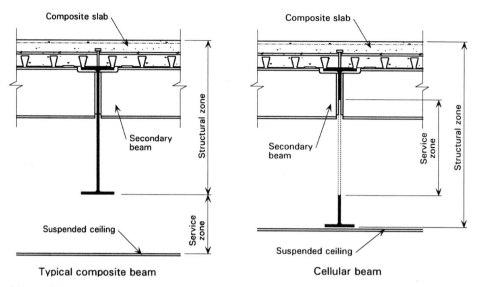

Typical composite beam   Cellular beam

Figure 4.31   Comparison between solid web beams and beams with web penetrations.

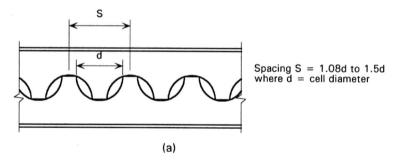

Spacing S = 1.08d to 1.5d
where d = cell diameter

(a)

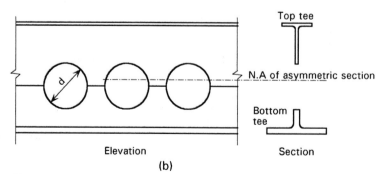

Top tee

N.A of asymmetric section

Bottom
tee

Elevation   Section
(b)

Figure 4.32   Cellular beam geometry. (a) Burning pofile. (b) Asymmetrical cellular beam.

design information is given in the SCI design guide *Design of composite and non-composite cellular beams.*[17]

The cellular beam is an efficient long-span member when considered as a simply supported beam subject to uniform loading. Point loads from secondary beams can be incorporated into the design, provided the secondary beams occur at a web post position. Also, the loads

Figure 4.33   Cellular beam cutting profile.

from the secondary beams should not be relatively high, as shear is the limiting factor in the design of the primary beams.

4.5.2 Benefits of cellular beam construction

Cellular beam construction produces an internal column-free environment at a similar building cost to short/medium-span structures using other forms of construction.

- *Accommodation of services*
  Floor zones are kept to a minimum by passing the services through the circular openings. This has the effect of reducing the overall height of the building when compared to the traditional approach of using a rolled section to form a composite beam (see Chapter 2 and Fig. 4.31). The reduced building height will provide savings in the cladding costs.

- *Asymmetric beam configuration*
  The cellular beam can be fabricated with a smaller top T-section when compared to the bottom T-section (see Fig. 4.32). This produces an efficient composite section (for use as a simply supported beam) when the upper smaller T-section is combined with the effective compression flange of the span. However, the top T-section should be designed for the forces developed during construction; this is covered by the cellbeam software.

- *Cambering*
  Figure 4.33 shows the cutting process for cellular beams. The web cutting process releases the residual stresses that are induced from the initial rolling process of the beam. As the two halves of the beam are separated, the T-sections develop a natural tendency to camber. The two halves of the beam are moved in to a position maintaining the natural camber and are welded. This method of cambering is free of charge to the user.

4.5.3 Beam geometry

The spacing of the cells (circular holes) can be arranged between the limits $1.08d$ to $1.6d$ (where d is the diameter of the cell). This provides flexibility in the positioning of the cells so that the secondary beam can coincide with the full web section. The advantage of using this technique means that fewer cells will require infilling at and around the secondary beam location.

It is possible to provide very large openings at strategic positions along the cellular beam to allow for the passage of unusually large ducts. However, the requirements for large ductwork should be kept to a minimum otherwise this will affect the economy of the system.

4.5.4 *Examples of use*

There are now many examples of the use of long-span composite cellular beams. Figures 4.34 and 4.35 illustrate good examples of recent projects. In both applications, the cellular beams had large openings through which circular air-conditioning ducts passed. Spans ranged from 12–18 m when used as long-span secondary beams.

Figure 4.34 Governor's House, London. Consulting Engineers: Ove Arup and Partners. Architect: Siddell Gibson.

Figure 4.35 ICMB Edinburgh University. Consulting Engineers: Ove Arup and Partners. Architect: Thomas Henny.

Appendix A
Composite Construction using Shallow Profiled Steel Decking

A1 Derivation of formulae used for composite construction with shallow profiled steel decking

The formulae shown in Appendix A1 have been used in BS 5950: Part 3: Section 3.1.

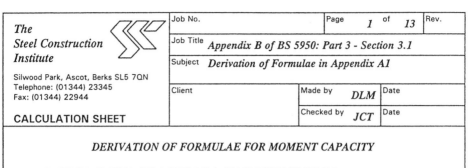

	Job No.		Page 1	of 13	Rev.
The Steel Construction Institute	Job Title *Appendix B of BS 5950: Part 3 - Section 3.1*				
	Subject *Derivation of Formulae in Appendix A1*				
Silwood Park, Ascot, Berks SL5 7QN Telephone: (01344) 23345 Fax: (01344) 22944	Client		Made by *DLM*	Date	
CALCULATION SHEET			Checked by *JCT*	Date	

DERIVATION OF FORMULAE FOR MOMENT CAPACITY

A.1 PLASTIC MOMENT CAPACITY OF A COMPOSITE SECTION

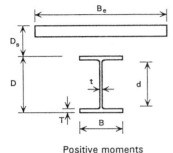

Positive moments   Negative moments

A.1.1   Resistances

The plastic moment capacity is expressed in terms of the resistance of various elements of the beam as follows:

Resistance of concrete flange,	R_c =	$0.45\,f_{cu}\,B_e\,(D_s - D_p)$
Resistance of steel flange,	R_f =	$B\,T\,p_y$
Resistance of slender steel beam,	R_n =	$R_s - R_v + R_o$
Resistance of slender web,	R_o =	$38\,\epsilon\,t^2\,p_y$
Resistance of shear connection,	R_q =	NQ
Resistance of reinforcement,	R_r =	$0.87\,f_y\,A_r$
Resistance of steel beam,	R_s =	$A\,p_y$
Resistance of clear web depth,	R_v =	$d\,t\,p_y$
Resistance of overall web depth,	R_w =	$R_s - 2\,R_f$

The *Steel Construction* *Institute* Silwood Park, Ascot, Berks SL5 7QN Telephone: (01344) 23345 Fax: (01344) 22944 **CALCULATION SHEET**		Job No.		Page *2* of *13*	Rev.

Job Title *Appendix B of BS 5950: Part 3 - Section 3.1*	
Subject *Derivation of Formulae in Appendix A1*	
Client	Made by *DLM* Date
	Checked by *JCT* Date

where: *A*   *is area of steel beam*

        A_r *is area of reinforcement*

        B   *is breadth of steel flange*

        B_e *is effective breadth of concrete flange*

        D_p *is depth of profiled steel sheet*

        D_s *is overall depth of concrete flange*

        d   *is clear depth of web*

        f_{cu} *is characteristic strength of concrete*

        f_y *is characteristic strength of reinforcement*

        N   *is number of shear connectors for positive or negative moment as relevant (minimum number one side of the point of maximum moment)*

        p_y *is design strength of steel (in N/mm^2)*

        Q   *is capacity of shear connectors for positive or negative moments as relevant*

        T   *is thickness of steel flange*

        t   *is thickness of web*

and ϵ *is constant* $(275/p_y)^{1/2}$

174   Composite Floor Systems

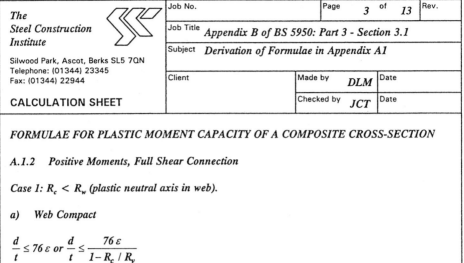

Job No.		Page	3	of	13	Rev.

The Steel Construction Institute

Silwood Park, Ascot, Berks SL5 7QN
Telephone: (01344) 23345
Fax: (01344) 22944

CALCULATION SHEET

Job Title *Appendix B of BS 5950: Part 3 - Section 3.1*

Subject *Derivation of Formulae in Appendix A1*

Client

Made by *DLM* Date

Checked by *JCT* Date

FORMULAE FOR PLASTIC MOMENT CAPACITY OF A COMPOSITE CROSS-SECTION

A.1.2   Positive Moments, Full Shear Connection

Case 1: $R_c < R_w$ *(plastic neutral axis in web).*

a)   Web Compact

$$\frac{d}{t} \le 76\,\varepsilon \quad or \quad \frac{d}{t} \le \frac{76\,\varepsilon}{1 - R_c / R_v}$$

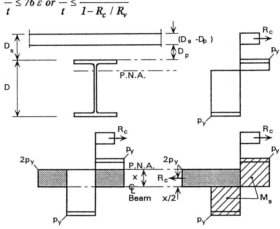

Moments about Centre Line of Beam

$$M_c = M_s + R_c \left[\frac{D}{2} + D_p + \frac{(D_s + D_p)}{2} \right] - R_c \cdot \frac{x}{2}$$

where $x = \dfrac{R_c}{t \times 2p_y} = \dfrac{R_c}{2R_v / d} = \dfrac{d}{2} \cdot \dfrac{R_c}{R_v}$ $R_v = d \cdot t \cdot p_y$

$$M_c = M_s + R_c \left[\frac{D + 2D_p + D_s - D_p}{2} \right] - R_c \frac{d}{2} \cdot \frac{R_c}{R_v} \cdot \frac{1}{2}$$

$$\boxed{M_c = M_s + R_c \left[\frac{D + D_s + D_p}{2} \right] - \frac{R_c^2}{R_v} \cdot \frac{d}{4}}$$

	Job No.		Page	4	of	13	Rev.
The Steel Construction Institute	Job Title	*Appendix B of BS 5950: Part 3 - Section 3.1*					
Silwood Park, Ascot, Berks SL5 7QN Telephone: (01344) 23345 Fax: (01344) 22944	Subject	*Derivation of Formulae in Appendix A1*					
	Client		Made by	*DLM*	Date		
CALCULATION SHEET			Checked by	*JCT*	Date		

b) *Web not compact*

$$\frac{d}{t} > \frac{76\,\varepsilon}{1 - R_c / R_v}$$

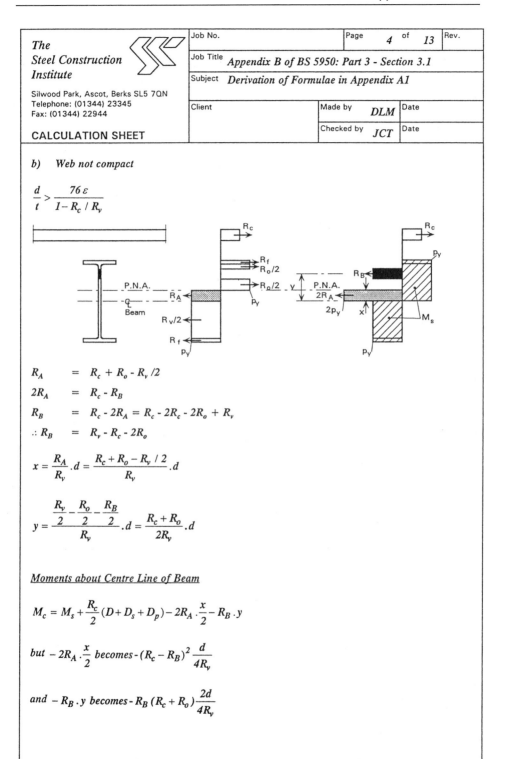

$$R_A = R_c + R_o - R_v/2$$

$$2R_A = R_c - R_B$$

$$R_B = R_c - 2R_A = R_c - 2R_c - 2R_o + R_v$$

$$\therefore R_B = R_v - R_c - 2R_o$$

$$x = \frac{R_A}{R_v}\cdot d = \frac{R_c + R_o - R_v/2}{R_v}\cdot d$$

$$y = \frac{\dfrac{R_v}{2} - \dfrac{R_o}{2} - \dfrac{R_B}{2}}{R_v}\cdot d = \frac{R_c + R_o}{2R_v}\cdot d$$

Moments about Centre Line of Beam

$$M_c = M_s + \frac{R_c}{2}(D + D_s + D_p) - 2R_A\cdot\frac{x}{2} - R_B\cdot y$$

$$but\ -2R_A\cdot\frac{x}{2}\ becomes - (R_c - R_B)^2\frac{d}{4R_v}$$

$$and\ -R_B\cdot y\ becomes - R_B\,(R_c + R_o)\frac{2d}{4R_v}$$

		Job No.		Page 5 of 13	Rev.
The *Steel Construction* *Institute*		Job Title *Appendix B of BS 5950: Part 3 - Section 3.1*			
		Subject *Derivation of Formulae in Appendix A1*			
Silwood Park, Ascot, Berks SL5 7QN Telephone: (01344) 23345 Fax: (01344) 22944		Client		Made by *DLM*	Date
CALCULATION SHEET				Checked by *JCT*	Date

Therefore,

$$-2R_A \cdot \frac{x}{2} - R_B \cdot y = -(R_c - R_B)^2 \frac{d}{4R_v} - R_B (R_c + R_o) \frac{2d}{4R_v}$$

$$= -\frac{d}{4R_v} [R_c^2 + R_B (R_B + 2R_o)]$$

but $R_B = R_v - R_c - 2R_o$

$$-2R_A \frac{x}{2} - R_B y = -\frac{d}{4R_v} [(R_v - R_c)(R_v - R_c - 2R_o) + R_c^2]$$

$$\boxed{M_c = M_s + \frac{R_c}{2}(D + D_s + D_p) - \frac{[R_c^2 + (R_v - R_c)(R_v - R_c - 2R_o)]}{R_v} \cdot \frac{d}{4}}$$

Case 2: $R_c \geq R_w$ (PNA in flange)

a) $R_s > R_c$ (PNA in steel flange)

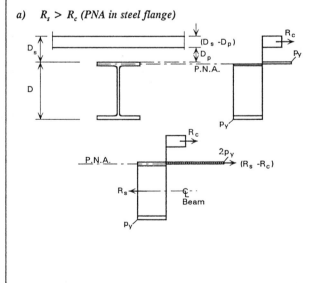

	Job No.		Page 6 of 13		Rev.
The **Steel Construction** **Institute**	Job Title *Appendix B of BS 5950: Part 3 - Section 3.1*				
	Subject *Derivation of Formulae in Appendix A1*				
Silwood Park, Ascot, Berks SL5 7QN Telephone: (01344) 23345 Fax: (01344) 22944	Client		Made by *DLM*	Date	
CALCULATION SHEET			Checked by *JCT*	Date	

Neutral axis depth, x

$$B \cdot x \cdot p_y + R_c = R_s - B \cdot x \cdot p_y$$

$$2B \cdot x \cdot p_y = R_s - R_c$$

$$\therefore \quad x = \frac{R_s - R_c}{2B \cdot p_y}$$

$$= \frac{R_s - R_c}{2R_f / T}$$

Moments about top flange of beam

$$M_c = R_s \cdot \frac{D}{2} + R_c \left(D_p + \frac{(D_s - D_p)}{2} \right) - (R_s - R_c) \frac{x}{2}$$

$$\boxed{M_c = R_s \cdot \frac{D}{2} + R_c \frac{(D_p + D_s)}{2} - \frac{(R_s - R_c)^2}{R_f} \cdot \frac{T}{4}}$$

Case 2 b) $R_s \le R_c$ *(PNA in Concrete flange)*

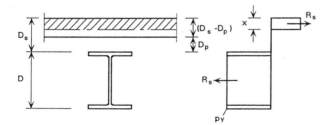

Neutral axis depth, x

$$x = \frac{R_s}{R_c} (D_s - D_p)$$

	Job No.		Page	7	of	13	Rev.

The
Steel Construction
Institute

Silwood Park, Ascot, Berks SL5 7QN
Telephone: (01344) 23345
Fax: (01344) 22944

CALCULATION SHEET

Job Title   *Appendix B of BS 5950: Part 3 - Section 3.1*

Subject   *Derivation of Formulae in Appendix A1*

Client

Made by   *DLM*   Date

Checked by   *JCT*   Date

Moments about top of concrete flange

$$M_c = R_s \left(D_s + \frac{D}{2} \right) - R_s . \frac{x}{2}$$

$$= R_s \left(D_s + \frac{D}{2} \right) - R_s . \frac{R_s}{2R_c} (D_s - D_p)$$

$$M_c = R_s . \left[\frac{D}{2} + D_s - \frac{R_s}{R_c} \frac{(D_s - D_p)}{2} \right]$$

A.1.3   Positive Moments, Partial Shear Connection

Case 3: $R_q < R_w$ (PNA in web).

a) $\frac{d}{t} \leq 76\,\varepsilon$ *or* $\frac{d}{t} \leq \frac{76\,\varepsilon}{1 - R_q / R_v}$ *(Web compact).*

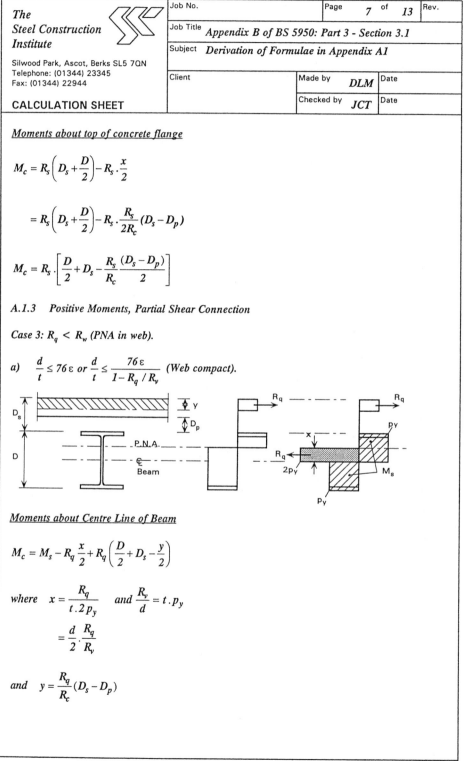

Moments about Centre Line of Beam

$$M_c = M_s - R_q \frac{x}{2} + R_q \left(\frac{D}{2} + D_s - \frac{y}{2} \right)$$

where $x = \dfrac{R_q}{t . 2 p_y}$ *and* $\dfrac{R_v}{d} = t . p_y$

$$= \frac{d}{2} . \frac{R_q}{R_v}$$

and $y = \dfrac{R_q}{R_c} (D_s - D_p)$

Job No.		Page 8 of 13	Rev.
Job Title *Appendix B of BS 5950: Part 3 - Section 3.1*			
Subject *Derivation of Formulae in Appendix A1*			
Client		Made by DLM	Date
		Checked by JCT	Date

The
Steel Construction
Institute

Silwood Park, Ascot, Berks SL5 7QN
Telephone: (01344) 23345
Fax: (01344) 22944

CALCULATION SHEET

$$M_c = M_s + R_q \left(\frac{D}{2} + D_s - \frac{R_q}{R_c} \cdot \frac{D_s - D_p}{2} \right) - \frac{R_q^2}{R_v} \cdot \frac{d}{4}$$

Case 3 (b) $\dfrac{d}{t} > \dfrac{76\,\varepsilon}{1 - R_q / R_v}$ *Web not compact.*

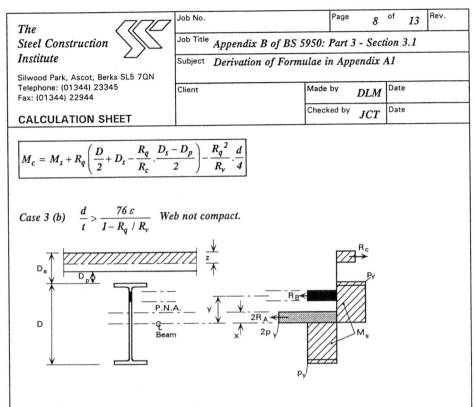

For build-up of diagrams see Case 1 (b)

$$R_A = R_q + R_o - R_v / 2$$

$$2R_A = R_q - R_B$$

$$R_B = R_v - R_q - 2R_o$$

$$x = \frac{R_A}{R_v} \cdot d$$

$$y = \frac{R_q + R_o}{2R_v} \cdot d$$

$$z = \frac{R_q}{R_c} (D_s - D_p)$$

$$M_c = M_s + R_q \left(\frac{D}{2} + D_s - \frac{z}{2} \right) - 2R_A \frac{x}{2} - R_B \cdot y$$

	Job No.			Page 9 of 13	Rev.
The **Steel Construction Institute**	Job Title *Appendix B of BS 5950: Part 3 - Section 3.1*				
Silwood Park, Ascot, Berks SL5 7QN Telephone: (01344) 23345 Fax: (01344) 22944	Subject *Derivation of Formulae in Appendix A1*				
	Client		Made by *DLM*	Date	
CALCULATION SHEET			Checked by *JCT*	Date	

where;

$$-2R_A \cdot \frac{x}{2} = -(R_q - R_B)^2 \frac{d}{4R_v}$$

$$-R_B \cdot y = -2R_B (R_q + R_o) \frac{d}{4R_v}$$

$$R_q \left(\frac{D}{2} + D_s - \frac{z}{2} \right) = R_q \left(\frac{D}{2} + D_s - \frac{R_q}{2R_c} (D_s - D_p) \right)$$

$$M_c = M_s + R_q \left[\frac{D}{2} + D_s - \frac{R_q}{R_c} \frac{(D_s - D_p)}{2} \right] - (R_q - R_B)^2 \frac{d}{4R_v} - 2R_B (R_q + R_o) \frac{d}{4R_v}$$

The last two terms are similar to Case 1(b) except that R_q replaces R_c

$$\boxed{M_c = M_s + R_q \left[\frac{D}{2} + D_s - \frac{R_q}{R_c} \frac{(D_s - D_p)}{2} \right] - \frac{R_q^2 + (R_v - R_q)(R_v - R_q + 2R_o)}{R_v} \cdot \frac{d}{4}}$$

Case 4 $R_q \geq R_w$ (PNA in flange)

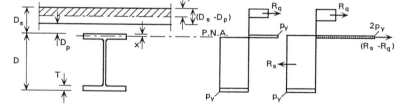

$$x = \frac{R_s - R_q}{2p_y B} = \frac{R_s - R_q}{2R_f / T}$$

$$y = \frac{R_q}{R_c} (D_s - D_p)$$

The Steel Construction Institute	Job No.					Page 10 of 13		Rev.
	Job Title *Appendix B of BS 5950: Part 3 - Section 3.1*							
Silwood Park, Ascot, Berks SL5 7QN	Subject *Derivation of Formulae in Appendix A1*							
Telephone: (01344) 23345 Fax: (01344) 22944	Client				Made by *DLM*		Date	
CALCULATION SHEET					Checked by *JCT*		Date	

Moments about top of steel flange

$$M_c = R_s \cdot \frac{D}{2} + R_q\left(D_s - \frac{y}{2}\right) - (R_s - R_q)\frac{x}{2}$$

$$\boxed{M_c = R_s \cdot \frac{D}{2} + R_q\left[D_s - \frac{R_q}{R_c}\frac{(D_s - D_p)}{2}\right] - \frac{(R_s - R_q)^2}{R_f} \cdot \frac{T}{4}}$$

A.1.4   Negative Moments

Case 5: Plastic neutral axis in Web

a) $R_r < R_w$.

$$\frac{d}{t} \le 38\varepsilon \ or \ \frac{d}{t} \le \frac{76\varepsilon}{1 - R_r / R_v} \quad (Web \ compact).$$

Moments about Centre Line of Beam

$$M_c = M_s + R_r\left[\frac{D}{2} + D_r\right] - R_r \cdot \frac{x}{2}$$

$$x = \frac{R_r}{2 p_y t} = \frac{R_r}{2R_v / d} = \frac{R_r}{R_v} \cdot \frac{d}{2}$$

$$\boxed{M_c = M_s + R_r\left[\frac{D}{2} + D_r\right] - \frac{R_r^2}{R_v}\frac{d}{4}}$$

	Job No.		Page _11_	of _13_	Rev.
The **Steel Construction** **Institute**	Job Title _Appendix B of BS 5950: Part 3 - Section 3.1_				
	Subject _Derivation of Formulae in Appendix A1_				
Silwood Park, Ascot, Berks SL5 7QN Telephone: (01344) 23345 Fax: (01344) 22944	Client		Made by _DLM_	Date	
CALCULATION SHEET			Checked by _JCT_	Date	

Case 5

b) $\dfrac{d}{t} > \dfrac{76\,\varepsilon}{1 - R_r / R_v}$ *(Web not compact)*

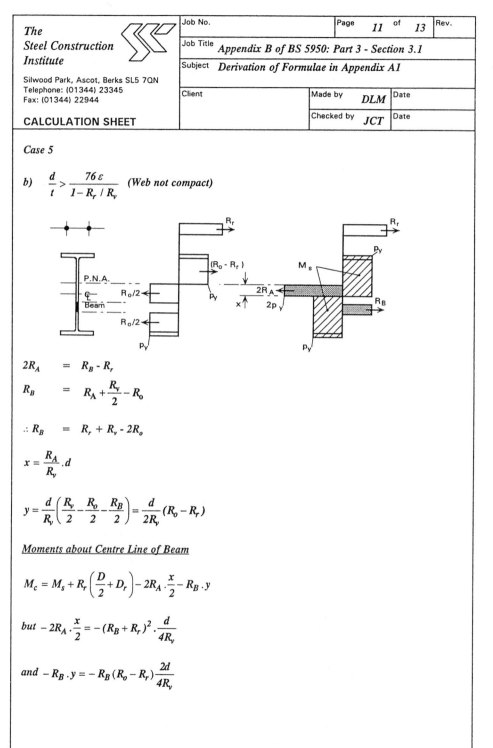

$2R_A = R_B - R_r$

$R_B = R_A + \dfrac{R_v}{2} - R_o$

$\therefore R_B = R_r + R_v - 2R_o$

$x = \dfrac{R_A}{R_v}\,.d$

$y = \dfrac{d}{R_v}\left(\dfrac{R_v}{2} - \dfrac{R_o}{2} - \dfrac{R_B}{2}\right) = \dfrac{d}{2R_v}(R_o - R_r)$

Moments about Centre Line of Beam

$M_c = M_s + R_r\left(\dfrac{D}{2} + D_r\right) - 2R_A\,.\dfrac{x}{2} - R_B\,.y$

$but\ -2R_A\,.\dfrac{x}{2} = -(R_B + R_r)^2\,.\dfrac{d}{4R_v}$

$and\ -R_B\,.y = -R_B(R_o - R_r)\dfrac{2d}{4R_v}$

The Steel Construction Institute	Job No.		Page 12 of 13	Rev.

Silwood Park, Ascot, Berks SL5 7QN
Telephone: (01344) 23345
Fax: (01344) 22944

CALCULATION SHEET

Job Title *Appendix B of BS 5950: Part 3 - Section 3.1*

Subject *Derivation of Formulae in Appendix A1*

Client

Made by **DLM** Date

Checked by **JCT** Date

$$\therefore -2R_A \cdot \frac{x}{2} - R_B \cdot y = -(R_B + R_r)^2 \frac{d}{4R_v} - R_B (R_o - R_r) \frac{2d}{4R_v}$$

$$= -\frac{d}{4R_v} (R_r^2 + R_B (R_B + 2R_o))$$

$$\text{but } R_B = R_r + R_v - 2R_o$$

$$2R_A \frac{x}{2} - R_B y = -\frac{d}{4R_v} [R_r^2 + (R_r + R_v - 2R_o)(R_r + R_v)]$$

$$\boxed{\therefore M_c = M_s + R_r \left(\frac{D}{2} + D_r \right) - \frac{d}{4R_v} [R_r^2 + (R_r + R_v - 2R_o)(R_r + R_v)]}$$

Case 6: Plastic neutral axis in flange

a) $\dfrac{d}{t} > 38\varepsilon$ *(Web compact)*

   $R_r \geqslant R_w$

i) $R_r < R_s$ *(PNA in steel flange)*

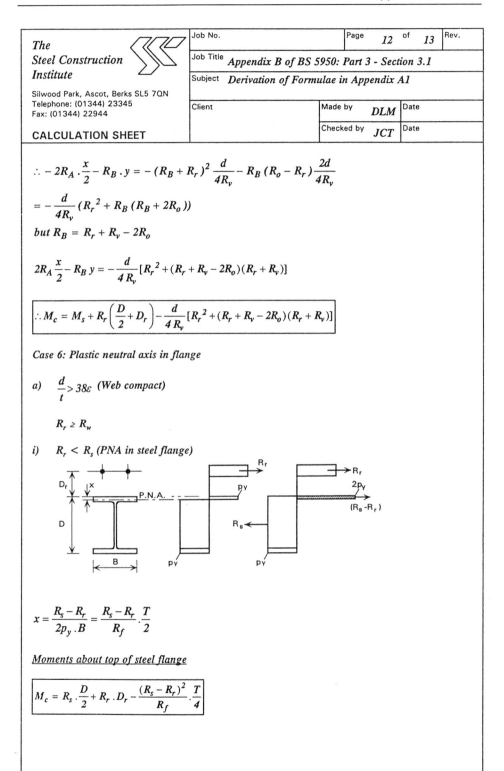

$$x = \frac{R_s - R_r}{2p_y \cdot B} = \frac{R_s - R_r}{R_f} \cdot \frac{T}{2}$$

<u>*Moments about top of steel flange*</u>

$$\boxed{M_c = R_s \cdot \frac{D}{2} + R_r \cdot D_r - \frac{(R_s - R_r)^2}{R_f} \cdot \frac{T}{4}}$$

	Job No.			Page 13	of 13	Rev.

The Steel Construction Institute

Silwood Park, Ascot, Berks SL5 7QN
Telephone: (01344) 23345
Fax: (01344) 22944

CALCULATION SHEET

Job Title *Appendix B of BS 5950: Part 3 - Section 3.1*

Subject *Derivation of Formulae in Appendix A1*

Client

Made by *DLM*   Date

Checked by *JCT*   Date

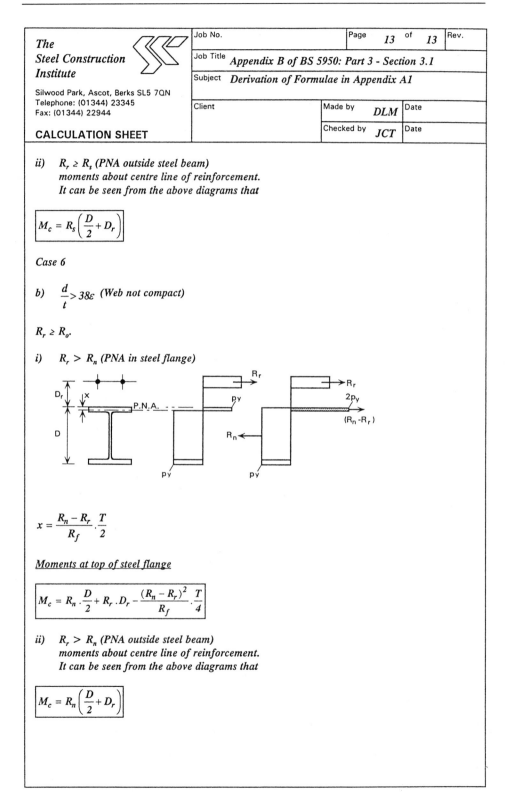

ii) $R_r \geq R_s$ *(PNA outside steel beam)*
   moments about centre line of reinforcement.
   It can be seen from the above diagrams that

$$M_c = R_s\left(\frac{D}{2} + D_r\right)$$

Case 6

b) $\dfrac{d}{t} > 38\varepsilon$ *(Web not compact)*

$R_r \geq R_o.$

i) $R_r > R_n$ *(PNA in steel flange)*

$$x = \frac{R_n - R_r}{R_f} \cdot \frac{T}{2}$$

Moments at top of steel flange

$$M_c = R_n \cdot \frac{D}{2} + R_r \cdot D_r - \frac{(R_n - R_r)^2}{R_f} \cdot \frac{T}{4}$$

ii) $R_r > R_n$ *(PNA outside steel beam)*
   moments about centre line of reinforcement.
   It can be seen from the above diagrams that

$$M_c = R_n\left(\frac{D}{2} + D_r\right)$$

A2 Worked example for a typical 10 m span composite beam subject to a uniformly distributed load

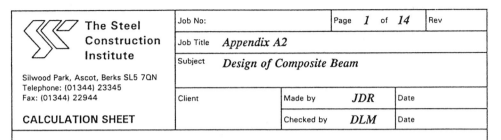

The Steel Construction Institute	Job No:	Page *1* of *14* Rev
	Job Title *Appendix A2*	
	Subject *Design of Composite Beam*	
Silwood Park, Ascot, Berks SL5 7QN Telephone: (01344) 23345 Fax: (01344) 22944	Client	Made by *JDR* Date
CALCULATION SHEET		Checked by *DLM* Date

DESIGN EXAMPLE

Consider internal composite beam A-A between columns, and subject to uniform loading.

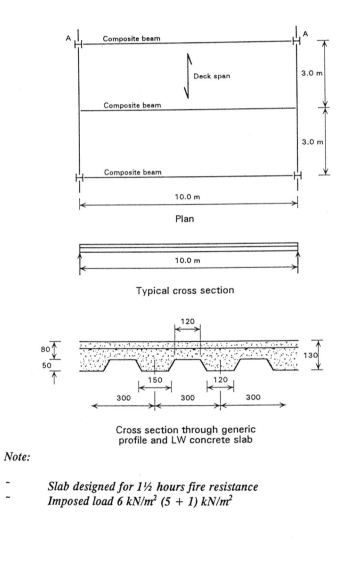

Plan

Typical cross section

Cross section through generic profile and LW concrete slab

Note:

- *Slab designed for 1½ hours fire resistance*
- *Imposed load 6 kN/m² (5 + 1) kN/m²*

The Steel Construction Institute	Job No:		Page **2** of **14**	Rev
	Job Title	*Appendix A2*		
Silwood Park, Ascot, Berks SL5 7QN	Subject	*Design of Composite Beam*		
Telephone: (01344) 23345 Fax: (01344) 22944	Client		Made by	*JDR*	Date
CALCULATION SHEET			Checked by	*DLM*	Date

DESIGN DATA
Floor Dimensions

Span	L	=	*10.0 m*
Beam spacing	b	=	*3.0 m*
Slab depth	D_s	=	*130 mm*

Deck: Generic Profile

Unpropped construction throughout

Shear Connectors   19 mm diameter
95 mm length after weld

MATERIALS

Steel: Grade S355
Design strength p_y =   355 N/mm² up to and including 16 mm thick
Concrete: Lightweight concrete grade 30
Density (dry) =   1800 kg/m³ (17.66 kN/m³)
Density (wet) =   1930 kg/m³ (18.93 kN/m³)

LOADING

Concrete slab

$$Dry\ weight = \left[130 \times 10^3 - \frac{50}{0.3}(120+30)\right]17.66/10^6$$

$$= 1.85\ kN/m^2$$

Wet weight = *1.85 × 18.93/17.66*
= *1.99 kN/m²*

Construction Stage			*kN/m²*
LW concrete slab (wet)		=	1.99
Steel deck		=	0.15
Reinforcement	allow	=	0.04
Steel beam	allow	=	0.20
			2.38 kN/m²
Construction load		=	0.50 kN/m²

Composite Stage		*kN/m²*
LW concrete slab (dry)	=	1.85
Steel deck	=	0.15

![SCI logo] **The Steel Construction Institute** Silwood Park, Ascot, Berks SL5 7QN Telephone: (01344) 23345 Fax: (01344) 22944	Job No:	Page **3** of **14**	Rev
	Job Title *Appendix A2*	
	Subject *Design of Composite Beam*	
	Client	Made by *JDR*	Date
CALCULATION SHEET	Checked by *DLM*	Date

Reinforcement   allow   =   0.04
Steel beam   allow   =   0.20
                                       2.24 kN/m²

Ceiling and services   =   0.50 kN/m²

	kN/m²
Imposed	
Occupancy	*5.0*
Partitions	*1.0*
	6.0 kN/m²

According to BS 6399, imposed loads may be reduced with respect to the beam supported area. For the purposes of this design example, this reduction will be omitted.

INITIAL SELECTION OF BEAM SIZE

Using Table 23 from reference 15, a suitable section for an imposed load of 6.0 kN/m² would be a 406 × 178 × 60 UB Grd S355

Section properties and dimensions

D	*=*	*406.4 mm*	*A =*	*76.0 cm²*
B	*=*	*177.8 mm*	I_x *=*	*21508 cm⁴*
t	*=*	*7.8 mm*	Z_x *=*	*1058 cm³*
T	*=*	*12.8 mm*	S_x *=*	*1194 cm³*
d	*=*	*360.5 mm*		

Design strength p_y = 355 N/mm² (12.8 < 16.0 mm)

CONSTRUCTION STAGE DESIGN
Ultimate Limit State Loading

Slab (wet) + beam   =   2.38 × 1.4   =   3.33
Construction   =   0.50 × 1.6   =   0.80
                                                        4.13 kN/m²

W =   4.13 × 10 × 3   =   123.9 kN
BM =   123.9 × 10/8   =   154.9 kNm

Assume the beam in the construction stage is laterally restrained by the decking.

∴ M_s = 1194 × 355/10³

The Steel Construction Institute	Job No:		Page *4* of *14*	Rev
	Job Title *Appendix A2*			
	Subject *Design of Composite Beam*			
Silwood Park, Ascot, Berks SL5 7QN Telephone: (01344) 23345 Fax: (01344) 22944	Client	Made by	*JDR*	Date
CALCULATION SHEET		Checked by	*DLM*	Date

$$= 423.9 \ kNm > 154.9 \ kNm \quad OK$$

$$S_x/Z_x \quad = 1194/1058 = \quad 1.13 < 1.2$$

∴ *Beam satisfactory for positive moment capacity in the construction stage*

COMPOSITE STAGE DESIGN
Ultimate Limit State Loading

			kN/m^2
Slab (dry) + beam	=	2.24 × 1.4 =	3.14
C&S	=	0.50 × 1.4 =	0.70
Imposed	=	6.0 × 1.6 =	9.60
			13.44 kN/m^2

$$W = 13.44 \times 10 \times 3 = 403.2 \ kN$$
$$BM = 403.2 \times 10/8 = 504.0 \ kNm$$

Plastic Positive Bending moment capacity
Check by using:

(a) *Linear interaction method}*   *See clause 2.2.6 of the text*
(b) *Stress block method}*   *See clause 2.2.6 of the text*

Also, for each method, check using 1 stud and 2 studs per trough

(a) *Linear Interaction Method to obtain moment capacity, M_c*

$$M_c = M_s + k \ (M_{pc} - M_s)$$

where: M_s = $S_x \, p_y = 1194 \times 355/10^3 = 423.9 \ kNm$
     k = *degree of shear connection*
     M_{pc} = *plastic moment capacity based on full shear connection*

Before M_c can be determined, calculate:

(i) *Shear connector capacity*

From Part 3.1 - Table 5 the shear connector strength = 100 kN

for LWC = 100 × 0.9 = 90 kN
∴ Q_p = 0.8 Q_k = 0.8 × 90 = 72 kN

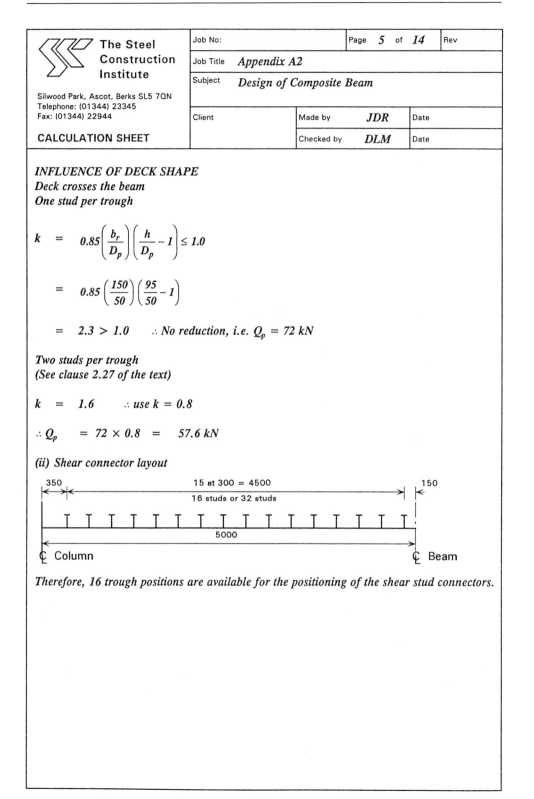

INFLUENCE OF DECK SHAPE
Deck crosses the beam
One stud per trough

$$k = 0.85 \left(\frac{b_r}{D_p} \right) \left(\frac{h}{D_p} - 1 \right) \leq 1.0$$

$$= 0.85 \left(\frac{150}{50} \right) \left(\frac{95}{50} - 1 \right)$$

$$= 2.3 > 1.0 \quad \therefore \text{No reduction, i.e. } Q_p = 72 \text{ kN}$$

Two studs per trough
(See clause 2.27 of the text)

$$k = 1.6 \quad \therefore \text{use } k = 0.8$$

$$\therefore Q_p = 72 \times 0.8 = 57.6 \text{ kN}$$

(ii) Shear connector layout

Therefore, 16 trough positions are available for the positioning of the shear stud connectors.

The Steel Construction Institute	Job No:		Page **6** of **14**	Rev
	Job Title *Appendix A2*			
Silwood Park, Ascot, Berks SL5 7QN Telephone: (01344) 23345 Fax: (01344) 22944	Subject *Design of Composite Beam*			
	Client	Made by *JDR*	Date	
CALCULATION SHEET		Checked by *DLM*	Date	

EFFECTIVE WIDTH OF COMPRESSION FLANGE

$B_e = 10 \times 10^3/4 = 2500\ mm < 3000\ mm\ beam\ crs.$

$\therefore B_e = 2500\ mm$

$R_c = 0.45 \times 2500 \times 80 \times 30/10^3 = 2700\ kN$
$R_s = 76 \times 355/10 = 2698\ kN$

$R_q\ (1\ stud) = 16 \times 72 = 1152\ kN$
$R_q\ (2\ studs) = 16 \times 2 \times 57.6 = 1843\ kN$

POSITIVE MOMENTS, FULL SHEAR CONNECTION

Case 2(b) ¯ pna in concrete flange ($R_s \le R_c$)

$$M_{pc} = R_s\left[\frac{D}{2} + D_s - \frac{R_s}{R_c}\left(\frac{D_s - D_p}{2}\right)\right] \quad (see\ Appendix\ A1)$$

$$= \frac{2698}{10^3}\left[\frac{406.4}{2} + 130 - \frac{2698}{2700} \times \frac{80}{2}\right]$$

$$= 791.1\ kNm$$

DEGREE OF SHEAR CONNECTION, k

$$k = R_q/R_s = \frac{1152}{2698} = 0.427\ (1\ stud/trough) > 0.4$$

$$\therefore M_c = 423.9 + 0.427\ (791.1 - 423.9)$$

$$= 580.7\ kNm$$

Using the linear interaction method with one stud per trough gives a positive moment capacity equal to:

580.7 kNm > 504 kNm   OK

		Job No:		Page 7 of 14		Rev
The Steel Construction Institute		Job Title	*Appendix A2*			
		Subject	*Design of Composite Beam*			
Silwood Park, Ascot, Berks SL5 7QN Telephone: (01344) 23345 Fax: (01344) 22944		Client		Made by	*JDR*	Date
CALCULATION SHEET				Checked by	*DLM*	Date

(b) Stress Block Method (1 stud/trough)

$$\frac{N_a}{N_p} \geq (L - 6)/10 \qquad but \geq 0.4$$

$$\therefore N_p = R_s/Q_p = \frac{2698}{72} = 37.5$$

$$\therefore N_a = 0.4 \times 37.5 = 15 \ but \ provide \ one \ stud \ per \ trough, \ i.e. \ N_a = 16$$

$$R_w = R_s - 2R_f$$

$$R_s = 2698 \ kN$$

$$R_f = BT \ p_y = 177.8 \times 12.8 \times 355/10^3 = 807.9 \ kN$$

$$\therefore R_w = 2698 - 2 \times 807.9$$

$$= 1082.2 \ kN$$

$$R_q = 1152 \ kN$$

$$\therefore R_q > R_w \ (pna \ in \ flange) \ \tilde{} \ Case \ 4$$

$$M_c = R_s . \frac{D}{2} + R_q \left[D_s - \frac{R_q}{R_c} \left(\frac{D_s - D_p}{2} \right) \right] - \left(\frac{R_s - R_q}{R_f} \right)^2 . \frac{T}{4}$$

$$R_f = 807.9 \ kN \ (see \ above)$$

$$M_c = \frac{2698}{10^3} \times \frac{406.4}{2} + \frac{1152}{10^3} \left[130 - \frac{1152}{2700} \times \frac{80}{2} \right] - \left(\frac{2698 - 1152}{807.9} \right)^2 \times \frac{12.8}{4 \times 10^3}$$

$$= 668.9 \ kNm$$

Using the stress block method with one stud per trough gives a positive moment capacity equal to:

668.9 kNm > 504 kNm   OK

	The Steel Construction Institute	Job No:		Page *8* of *14*	Rev
		Job Title	*Appendix A2*		
		Subject	*Design of Composite Beam*		

Silwood Park, Ascot, Berks SL5 7QN
Telephone: (01344) 23345
Fax: (01344) 22944

Client	Made by	*JDR*	Date
CALCULATION SHEET	Checked by	*DLM*	Date

USING 2 STUDS PER TROUGH

The procedures to obtain the moment capacity for 2 studs per trough are similar to the procedures when using 1 stud/trough. A shortened version of the calculations are as follows:

(a) Linear Interaction Method

$$M_{pc} = 791.1 \ kNm$$
$$R_q = 1843 \ kN$$
$$R_q/R_s = 1843/2698 = 0.68$$

$$\therefore M_c = 423.9 + 0.68 \ (791.1 - 423.9)$$
$$= 673.6 \ kNm$$

(b) Stress Block Method

$$N_p = 2698/57.6 = 47$$
$$N_a \ (min) = 0.4 \times 47 = 19 \ but \ 32 \ provided$$
$$\therefore R_q = 1843 \ kN \ (as \ above)$$

$$R_q > R_w \ ~ Case \ 4$$

$$M_c = \frac{2698}{10^3} \times \frac{406.4}{2} + \left[130 - \frac{1843}{2700} \times \frac{80}{2}\right]\frac{1843}{10^3} - \left(\frac{2698 - 1843}{807.9}\right)^2 \times \frac{12.8}{4}$$

$$= 733.9 \ kNm$$

SUMMARY

	Moment Capacity		Design Factored Moment
Studs per Trough	Linear Interaction	Stress Block	
1	*580.7*	*668.9*	*> 504*
2	*673.6*	*733.9*	

Moments in kNm

	Job No:		Page *9* of *14*	Rev
The Steel Construction Institute Silwood Park, Ascot, Berks SL5 7QN Telephone: (01344) 23345 Fax: (01344) 22944	Job Title	*Appendix A2*		
	Subject	*Design of Composite Beam*		
	Client	Made by	*JDR*	Date
CALCULATION SHEET		Checked by	*DLM*	Date

VERTICAL SHEAR

Construction stage reaction   = $2.24 \times 1.4 \times 10 \times 3/2$ =   *47 kN*

Composite stage reaction   = $[(6 \times 1.6) + (0.5 \times 1.4)] \, 10 \times 3/2$ =   *155 kN*

                                        Total   =   *202 kN*

SHEAR CAPACITY, P_v

P_v = $0.6 \, p_y \, A_v$ *where:* A_v =   *Dt*

    = $\dfrac{0.6 \times 355 \times 406.4 \times 7.8}{10^3}$ =   *675 kN*

        $0.5R$ = $338 \, kN > 202 \, kN$ *OK for all load cases*

N.B.   *With a uniformly distributed load it is not likely that shear would have an influence on moment capacity.*

SERVICEABILITY LIMIT STATES
Irreversible Deformation
Elastic Stresses in the Construction Stage

Slab + beam   = $2.24 \, kN/m^2$

W = $2.24 \times 10 \times 3$ =   *67.2 kN*

BM = $67.2 \times 10/8$ =   *84.0 kNm*

f_{steel} = $\dfrac{84 \times 10^6}{1058 \times 10^3}$ = $79.4 \, N/mm^2$

Composite Elastic Section Properties

Position of the e.n.a. from the upper surface of the slab.

y_e = $\dfrac{(D_s - D_p)0.5 + \alpha_e . r(D/2 + D_s)}{(1 + \alpha_e . r)}$

where: r = $A/[(D_s - D_p) \, B_e]$

		Job No:		Page *10* of *14*	Rev
The Steel Construction Institute		Job Title	*Appendix A2*		
		Subject	*Design of Composite Beam*		
Silwood Park, Ascot, Berks SL5 7QN Telephone: (01344) 23345 Fax: (01344) 22944		Client		Made by *JDR*	Date
CALCULATION SHEET				Checked by *DLM*	Date

α = 15 (average modular ratio for LWC)

r = $\dfrac{76 \times 10^2}{(130 - 50)\,2500}$ = 0.038

y_e = $\dfrac{80 \times 0.5 + 15 \times 0.038}{1 + 15 \times 0.038}\left(\dfrac{406.4}{2} + 130\right)$

   = 146.4 mm

UNCRACKED INERTIA, I_c

I_c = $\dfrac{A}{4}\left(\dfrac{D + D_s + D_p}{1 + \alpha_e \cdot r}\right)^2 + \dfrac{B_e\left(D_s - D_p\right)^3}{12\alpha_e} + I$

   = $\dfrac{76\,(40.64 + 13 + 5)^2}{4\,(1 + 15 \times 0.038)} + \dfrac{250 \times 8^3}{12 \times 15} + 21508$

   = 63833 cm⁴

Z_t (steel) = $I_c/(D + D_s - y_e)$

   = $\dfrac{63833}{40.64 + 13 - 14.64}$

   = 1636.7 cm³

Z_c (conc) = $I_c \cdot \alpha_e/y_e$

   = $\dfrac{63833 \times 15}{14.64}$

   = 65403 cm³

COMPOSITE STAGE LOADING

Imposed = 6.0 kN/m²
C&S   = 0.5 kN/m²
     6.5 kN/m²

The Steel Construction Institute Silwood Park, Ascot, Berks SL5 7QN Telephone: (01344) 23345 Fax: (01344) 22944 **CALCULATION SHEET**	Job No:	Page *11* of *14* Rev
	Job Title *Appendix A2*	
	Subject *Design of Composite Beam*	
	Client	Made by *JDR* Date
		Checked by *DLM* Date

W = $6.5 \times 10 \times 3$ = $195 \ kN$
BM = $195 \times 10/8$ = $244 \ kNm$
Extreme fibre stress ˜ Tension flange

f_{steel} = $\dfrac{244 \times 10^6}{1636.7 \times 10^3}$ = $149.1 \ N/mm^2$

Combined stress = $79.4 + 149.1 \ N/mm^2$

= $228.5 \ N/mm^2$ < p_y = $355 \ N/mm^2$

f_{conc} = $\dfrac{244 \times 10^6}{65403 \times 10^3}$ = $3.7 \ N/mm^2 < 0.5 f_{cu}$ = $15 \ N/mm^2$

∴ *Serviceability stresses satisfactory*

DEFLECTION
Construction Stage

w = $2.38 \ kN/m^2 \ (slab + beam)$
W = $2.38 \times 10 \times 3$ = $71.4 \ kN$

δ = $\dfrac{5 \times 71.4 \left(10 \times 10^3\right)^3}{384 \times 205 \times 21508 \times 10^4}$

= $21.1 \ mm \ (1/474)$

COMPOSITE STAGE

w = $6.0 \ kN/m^2 \ (Imposed)$
W = $6 \times 10 \times 3$ = $180 \ kN$

Deflection ˜ full shear connection

δ_c = $\dfrac{5 \times 180 \left(10 \times 10^3\right)^3}{384 \times 205 \times 63833 \times 10^4}$

= $17.9 \ mm$

		Job No:		Page *12* of *14*	Rev
The Steel Construction Institute		Job Title *Appendix A2*			
		Subject *Design of Composite Beam*			
Silwood Park, Ascot, Berks SL5 7QN Telephone: (01344) 23345 Fax: (01344) 22944		Client	Made by	*JDR*	Date
CALCULATION SHEET			Checked by	*DLM*	Date

In both of the cases, partial shear connection exists. The following equation takes into account the effects of slip.

$$\delta'_c \;=\; \delta_c + 0.3\,(1 - k)\,(\delta_o - \delta_c)$$

Degree of shear connection k (1 stud/trough) $=$ *0.427*

Degree of shear connection k (2 studs/trough) $=$ *0.68*

Deflection when k = 0.427

δ_o $=$ *Deflection for the steel beam acting alone*

$$= \;\; 17.9 \times \frac{63833}{21508} \;\; = \;\; 53.2 \; mm$$

$\therefore \delta'_c =$ *17.9 + 0.3 (1 - 0.427) (53.2 - 17.9)*

$$= \;\; 24.0 \; mm \;\; (1/417) < (1/360) \quad OK$$

$\therefore$ *Imposed deflection satisfactory for both cases*

TOTAL DEFLECTION (k = 0.427)

Construction stage $=$ *21.1 mm*

Imposed $=$ *24.0 mm*

$C\&S \approx \dfrac{0.5}{6.0} \times 24$ $=$ *2.0 mm*

                       47.1 mm

Pre-cambering would not be considered for a construction stage deflection of 21.1 mm over a span of 10.0 m.

TRANSVERSE REINFORCEMENT

Mesh ¯ A142
Check resistance of concrete flange

	Job No:		Page *13* of *14*	Rev
The Steel Construction Institute	Job Title	*Appendix A2*		
	Subject	*Design of Composite Beam*		
Silwood Park, Ascot, Berks SL5 7QN Telephone: (01344) 23345 Fax: (01344) 22944	Client		Made by	*JDR*	Date
CALCULATION SHEET			Checked by	*DLM*	Date

SHEAR RESISTANCE PER SHEAR SURFACE

$$V_r = 0.7\, A_{sv}.f_y + 0.03\, \eta.A_{cv}.f_{cu} + v_p$$

$$\text{but} \not> 0.8\eta.A_{cv}\, (f_{cu})^{½} + v_p$$

SHEAR FORCE PER UNIT LENGTH, v

2 ~ Shear connectors per trough

$$V = \frac{2 \times 57.6}{0.3} \times 0.5$$

$$= 192\ kN/m \text{ per shear plane}$$

$$A_{sv} = 142\ mm^2/m$$

$$A_{cv} = 105 \times 10^3\ mm^2/m$$

$$f_y = 460\ N/mm^2$$

$$\eta = 0.8\ for\ LWC$$

$$f_{cu} = 30\ N/mm^2\ LWC$$

$$v_p = t_p.p_{yb} = 0.9 \times 280 = 252\ kN/m$$

$$v_r = \frac{0.7 \times 142 \times 460}{10^3} + \frac{0.03 \times 0.8 \times 105 \times 10^3 \times 30}{10^3} + 252$$

$$= 373.3\ kN/m$$

$$0.8\ \eta.A_{cv}\, (f_{cu})^{½} + v_p = \frac{0.8 \times 0.8 \times 105 \times 10^3}{10^3}(30)^{1/2} + 252$$

$$= 620\ kN/m > 373.3\ kN/m$$

$$\therefore v_r = 373.3\ kN/m > 192\ kN/m \quad OK$$

A142 mesh satisfactory

The Steel Construction Institute	Job No:		Page *14* of *14*	Rev
	Job Title	*Appendix A2*		
Silwood Park, Ascot, Berks SL5 7QN	Subject	*Design of Composite Beam*		
Telephone: (01344) 23345 Fax: (01344) 22944	Client		Made by *JDR*	Date
CALCULATION SHEET			Checked by *DLM*	Date

VIBRATION
Simplified Approach

Loading kN/m^2

Slab + beam   2.24
C&S   0.50
10% occupancy   0.50
                   3.24 kN/m^2

W = $3.24 \times 10 \times 3$

   =   *97.2 kN*

Increase the inertia by 10% to allow for the increased dynamic stiffness of the composite beam.

= 63833×1.1 = $70216 \ cm^4$

δ = $\dfrac{5 \times 97.2 \times \left(10 \times 10^3\right)^3}{384 \times 205 \times 70216 \times 10^4}$ =   *8.8 mm*

Natural frequency $\approx \dfrac{18}{(8.8)^{1/2}}$ =   *6.0 Hz*

∴ *6 Hz > 4.0 Hz beam satisfactory for vibration*

N.B.   *For further information regarding this topic, see reference (16).*

Appendix B
Slimdek Construction

B1 Derivation of formulae for the Slimflor (SFB) beam

B1.1 Derivation of design formulae for the Slimflor beam (SFB)

	Job No:		Page *1* of *12*		Rev
The Steel Construction Institute	Job Title	*Appendix B1.1 - Elastic Section Properties*			
	Subject	*Slimflor Beam*			
Silwood Park, Ascot, Berks SL5 7QN Telephone: (01344) 23345 Fax: (01344) 22944	Client		Made by	*DLM*	Date
CALCULATION SHEET			Checked by	*JWR*	Date

(i)   *Steel section only*
(ii)   *Steel section plus concrete*

(i)   *STEEL SECTION ONLY*

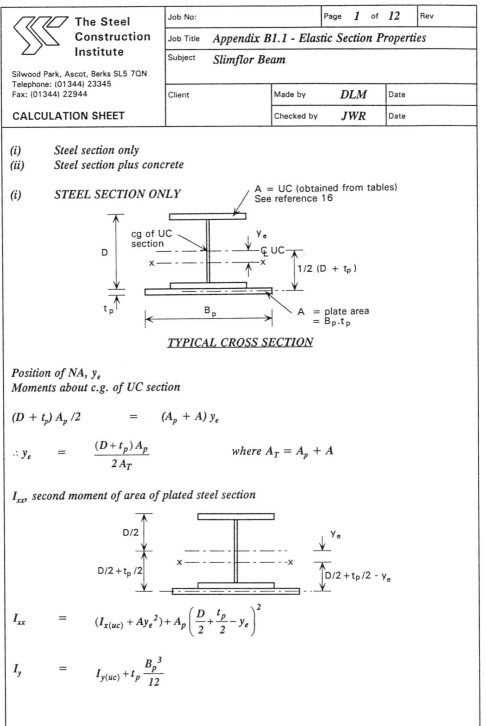

TYPICAL CROSS SECTION

Position of NA, y_e
Moments about c.g. of UC section

$$(D + t_p) A_p /2 \quad = \quad (A_p + A) y_e$$

$$\therefore y_e \quad = \quad \frac{(D + t_p) A_p}{2 A_T} \qquad \text{where } A_T = A_p + A$$

I_{xx}, second moment of area of plated steel section

$$I_{xx} \quad = \quad (I_{x(uc)} + A y_e^2) + A_p \left(\frac{D}{2} + \frac{t_p}{2} - y_e \right)^2$$

$$I_y \quad = \quad I_{y(uc)} + t_p \frac{B_p^3}{12}$$

![SCI logo] The Steel Construction Institute	Job No:		Page *2* of *12*	Rev
	Job Title *Appendix B1.1 - Elastic Section Properties*			
Silwood Park, Ascot, Berks SL5 7QN Telephone: (01344) 23345 Fax: (01344) 22944	Subject *Slimflor Beam*			
CALCULATION SHEET	Client	Made by *DLM*		Date
		Checked by *JWR*		Date

SECTION MODULI, Z_x

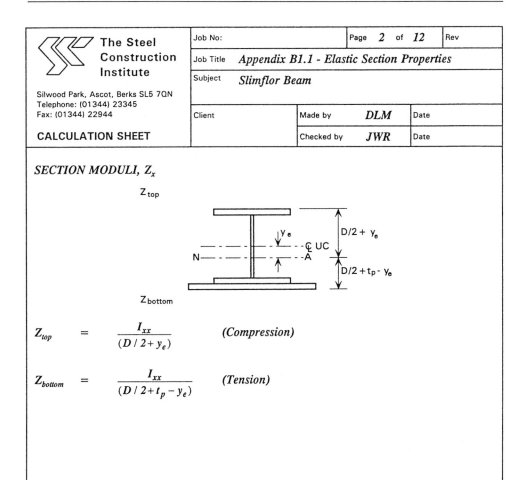

$$Z_{top} = \frac{I_{xx}}{(D/2 + y_e)} \qquad \textit{(Compression)}$$

$$Z_{bottom} = \frac{I_{xx}}{(D/2 + t_p - y_e)} \qquad \textit{(Tension)}$$

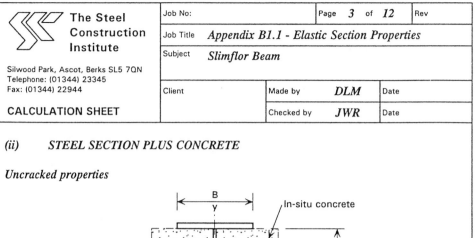

		Page **3** of **12**	Rev

The Steel Construction Institute

Silwood Park, Ascot, Berks SL5 7QN
Telephone: (01344) 23345
Fax: (01344) 22944

CALCULATION SHEET

Job No:			Page **3** of **12**	Rev
Job Title	*Appendix B1.1 - Elastic Section Properties*			
Subject	*Slimflor Beam*			
Client		Made by	*DLM*	Date
		Checked by	*JWR*	Date

(ii)   **STEEL SECTION PLUS CONCRETE**

Uncracked properties

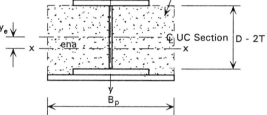

modular ratio, α_e

Concrete Area, A_c *(steel units)*

$$A_c = \frac{B_p(D-2T)}{\alpha_e}$$

Concrete second moment of area, I_{cx}

$$I_{cx} = \frac{B_p(D-2T)^3}{12\alpha_e} \qquad\qquad I_{cy} = \frac{(D-2T)B_p^{\ 3}}{12\alpha_e}$$

$$I_x = (I_{x(uc)} + Ay_c^{\ 2}) + A_p\left(\frac{D+t_p}{2} - y_c\right)^2 + B_p\frac{(D-2T)^3}{12\alpha_e} + A_c y_c^{\ 2}$$

$$I_{u\#y} = I_{y(uc)} + \frac{t_p B_p^{\ 3}}{12} + \frac{(D-2T)B_p^{\ 3}}{12\alpha_e}$$

where:

$$\left(\frac{D}{2} + \frac{t_p}{2}\right)A_p = (A_T + A_c)y_e$$

$$\therefore y_e = \frac{A_p(D+t_p)}{2(A_T + A_c)}$$

	The Steel Construction Institute	Job No:		Page *4* of *12*	Rev
		Job Title	*Appendix B1.1*		
		Subject	*Plastic Moment Capacity*		
Silwood Park, Ascot, Berks SL5 7QN Telephone: (01344) 23345 Fax: (01344) 22944		Client		Made by *DLM*	Date
CALCULATION SHEET				Checked by *JWR*	Date

PLASTIC MOMENT CAPACITY

Resistances

The plastic moment capacity is expressed in terms of the resistance of various elements of the beam as follows.

Resistance of concrete flange

$$R_c = 0.45 f_{cu} B_e D_s \qquad\qquad\qquad if\ D_d \geq D$$
$$R_c = 0.45 f_{cu} B_e (D_s + D_d - D + T) \qquad if\ D > D_d$$

Resistance of steel flange,	R_f	$=$	$B\,T\,p_y$
Resistance of flange plate,	R_p	$=$	$B_p\,t_p\,p_y$
Resistance of shear connection,	R_q	$=$	NQ
Resistance of steel beam,	R_s	$=$	$A\,p_y$
Resistance of clear web depth,	R_v	$=$	$d\,t_w\,p_y$
Resistance of overall web flange,	R_w	$=$	R_s - $2\,R_f$

where:

A *is the area of the steel beam;*
A_p *is the area of the flange plate;*
B *is the breadth of the steel flange;*
B_e *is the effective breadth of the concrete flange;*
B_p *is the breadth of the flange plate;*
D_d *is the depth of the steel decking;*
D_S *is the depth of the in situ concrete;*
d *is the clear depth of the web;*
f_{cu} *is the characteristic strength of the concrete;*
Q *is the capacity of a shear connector;*
T *is the flange thickness;*
t_w *is the web thickness;*
t_p *is the bottom flange plate thickness;*
S_x *plastic section moduli of the UC section;*

N is the actual number of shear connectors for positive moments as relevant (minimum number, one side of the point of maximum moment);

		Job No:			Page	5	of	12	Rev

The Steel Construction Institute

Silwood Park, Ascot, Berks SL5 7QN
Telephone: (01344) 23345
Fax: (01344) 22944

CALCULATION SHEET

Job Title	*Appendix B1.1*
Subject	*Plastic Moment Capacity*

Client		Made by	*DLM*	Date
		Checked by	*JWR*	Date

Case 1 SP (Steel Plastic): pna in web

Where $R_w > R_p$

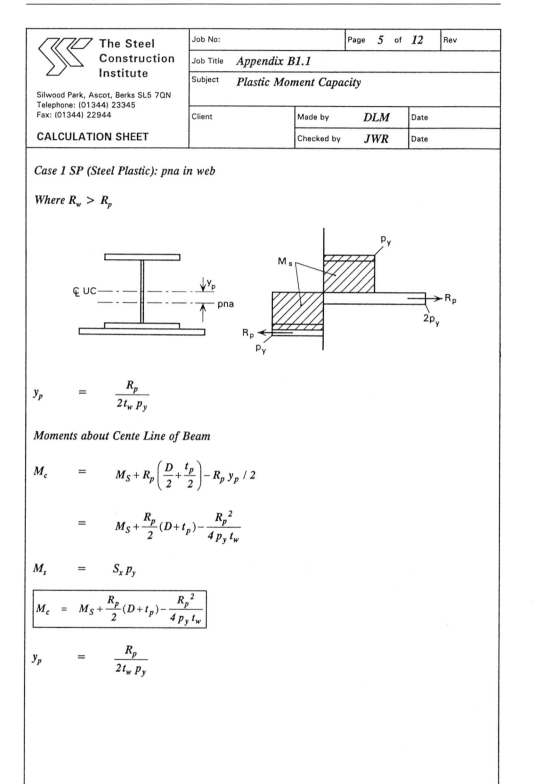

$$y_p = \frac{R_p}{2 t_w \, p_y}$$

Moments about Cente Line of Beam

$$M_c = M_S + R_p \left(\frac{D}{2} + \frac{t_p}{2} \right) - R_p \, y_p \,/\, 2$$

$$= M_S + \frac{R_p}{2}(D + t_p) - \frac{R_p^2}{4 \, p_y \, t_w}$$

$$M_s = S_x \, p_y$$

$$\boxed{M_c = M_S + \frac{R_p}{2}(D + t_p) - \frac{R_p^2}{4 \, p_y \, t_w}}$$

$$y_p = \frac{R_p}{2 t_w \, p_y}$$

		Job No:		Page	6	of	12	Rev

The Steel Construction Institute

Silwood Park, Ascot, Berks SL5 7QN
Telephone: (01344) 23345
Fax: (01344) 22944

CALCULATION SHEET

Job Title	*Appendix B1.1*
Subject	*Plastic Moment Capacity*

Client		Made by	*DLM*	Date	
		Checked by	*JWR*	Date	

Case 2 SP (Steel Plastic): pna in the UC flange

Where $(R_p \geq R_w)$ & $(R_s > R_p)$

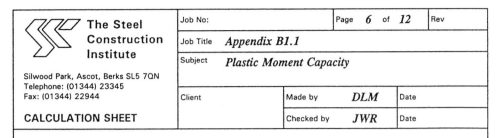

$$M_c = R_s D/2 + R_p t_p/2 - (R_s - R_p) y_p/2$$

where: $$y_p = \frac{(R_s - R_p)}{2 B p_y}$$

$$\therefore M_c = R_s D/2 + R_p t_p/2 - \frac{(R_s - R_p)^2}{4 B p_y}$$

$$\boxed{M_c = R_s \frac{D}{2} + R_p \frac{t_p}{2} - \frac{(R_s - R_p)^2}{4 B p_y}}$$

$$y_p = \left(\frac{R_s - R_p}{2 B p_y}\right)$$

	The Steel Construction Institute	Job No:		Page *7* of *12*		Rev
		Job Title	*Appendix B1.1*			
Silwood Park, Ascot, Berks SL5 7QN		Subject	*Plastic Moment Capacity*			
Telephone: (01344) 23345						
Fax: (01344) 22944		Client		Made by	*DLM*	Date
CALCULATION SHEET				Checked by	*JWR*	Date

Case 3 SP (Steel Plastic): pna in the bottom flange plate

$R_p > R_w$ & $R_p > R_s$

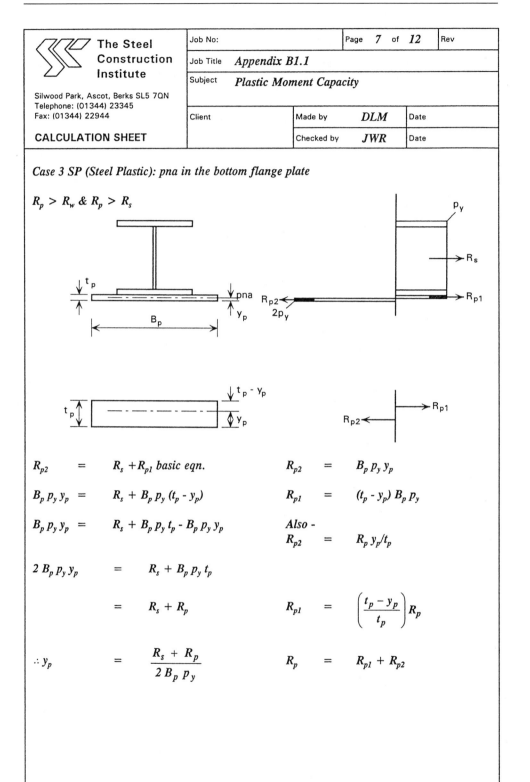

$$R_{p2} = R_s + R_{p1} \text{ basic eqn.} \qquad\qquad R_{p2} = B_p p_y y_p$$

$$B_p p_y y_p = R_s + B_p p_y (t_p - y_p) \qquad\qquad R_{p1} = (t_p - y_p) B_p p_y$$

$$B_p p_y y_p = R_s + B_p p_y t_p - B_p p_y y_p \qquad\qquad \text{Also -}$$

$$\qquad\qquad\qquad\qquad\qquad\qquad\qquad\qquad R_{p2} = R_p y_p / t_p$$

$$2 B_p p_y y_p = R_s + B_p p_y t_p$$

$$\qquad\qquad = R_s + R_p \qquad\qquad R_{p1} = \left(\frac{t_p - y_p}{t_p} \right) R_p$$

$$\therefore y_p = \frac{R_s + R_p}{2 B_p p_y} \qquad\qquad R_p = R_{p1} + R_{p2}$$

The Steel Construction Institute	Job No:	Page **8** of **12**	Rev
	Job Title *Appendix B1.1*		
	Subject *Plastic Moment Capacity*		

Silwood Park, Ascot, Berks SL5 7QN
Telephone: (01344) 23345
Fax: (01344) 22944

Client	Made by *DLM*	Date

CALCULATION SHEET

Checked by *JWR*	Date

$$M_c \quad = \quad R_{p2}\, y_p\,/\,2 + \frac{R_{p1}}{2}(t_p - y_p) + R_s\left(\frac{D}{2} + t_p - y_p\right)$$

$$= \quad \left(\frac{R_p\, y_p}{t_p}\right)\frac{y_p}{2} + \frac{R_p}{2}\left(\frac{t_p - y_p}{t_p}\right)(t_p - y_p) + R_s\left(\frac{D}{2} + t_p - y_p\right)$$

$$= \quad \frac{R_p}{2t_p}\left[y_p{}^2 + (t_p - y_p)^2\right] + R_s\,(D/2 + t_p - y_p)$$

$$\boxed{M_c \quad = \quad \frac{R_p}{2t_p}\left[y_p{}^2 + (t_p - y_p)^2\right] + R_s\,(D/2 + t_p - y_p)}$$

$$y_p \quad = \quad \frac{R_s + R_p}{2\,B_p\, p_y}$$

The Steel Construction Institute	Job No:	Page **9** of **12**	Rev
	Job Title *Appendix B1.1 - Plastic Moment Capacity*		
	Subject *Composite Section Based on Partial Shear Connection*		
Silwood Park, Ascot, Berks SL5 7QN Telephone: (01344) 23345 Fax: (01344) 22944			
	Client	Made by **DLM**	Date
CALCULATION SHEET		Checked by **JWR**	Date

CASE 1 PS (Partial Shear): pna in top flange of UC

$(R_p + R_s) > R_q > (R_p + R_w)$

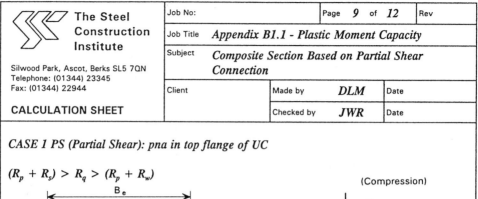

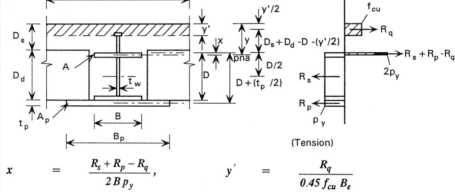

$$ x = \frac{R_s + R_p - R_q}{2 B p_y}, \qquad\qquad y' = \frac{R_q}{0.45 f_{cu} B_e} $$

Moments about top flange (UC)

$$ M_c = R_q (D_s + D_d - D - y'/2) + \frac{R_s D}{2} + R_p (D + t_p / 2) - (R_s + R_p - R_q)\frac{x}{2} $$

$$ \boxed{ M_c = R_q \left(D_s + D_d - D - \frac{R_q}{0.9 f_{cu} B_e} \right) + \frac{R_s D}{2} + R_p (D + t_p / 2) - \frac{(R_s + R_p - R_q)^2}{4 B p_y} } $$

Max value of y'

$$ y' = D_s \text{ when } D_d \geq D $$

$$ y' = (D_s + D_d - D + T) \text{ when } D > D_d $$

		Job No:		Page *10* of *12*		Rev
The Steel Construction Institute		Job Title	*Appendix B1.1 - Plastic Moment Capacity*			
		Subject	*Composite Section Based on Partial Shear Connection*			
Silwood Park, Ascot, Berks SL5 7QN Telephone: (01344) 23345 Fax: (01344) 22944						
CALCULATION SHEET		Client		Made by	*DLM*	Date
				Checked by	*JWR*	Date

CASE 2 PS (Partial Shear): pna in web of beam, above Centre Line of UC section

$(R_p + R_w) > R_q > R_p$

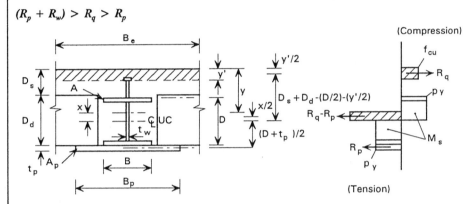

Moments about Centre Line (UC)

$$x = \frac{R_q - R_p}{2 p_y t_w}, \qquad y' = \frac{R_q}{0.45 f_{cu} B_e}$$

$$M_c = M_s + R_q (D_d + D_s - D/2 - y'/2) + \frac{R_p}{2}(D + t_p) - (R_q - R_p)\frac{x}{2}$$

$$\boxed{M_c = M_s + R_q \left(D_d + D_s - \frac{D}{2} - \frac{R_q}{0.9 f_{cu} B_e} \right) + \frac{R_q}{2}(D + t_p) - \frac{(R_q - R_p)^2}{4 p_y t_w}}$$

Max value y'

$y' = D_s$ when $D_d \geq D$

$y' = (D_s + D_d - D + T)$ when $D > D_d$

	Job No:		Page *11* of *12*	Rev
The Steel Construction Institute	Job Title	*Appendix B1.1 - Plastic Moment Capacity*		
	Subject	*Composite Section Based on Partial Shear Connection*		
Silwood Park, Ascot, Berks SL5 7QN Telephone: (01344) 23345 Fax: (01344) 22944				
CALCULATION SHEET	Client		Made by *DLM*	Date
			Checked by *JWR*	Date

CASE 3 PS (Partial Shear): pna within the web of the UC but below the UC Centre Line

$$R_p > R_q > (R_p - R_w)$$

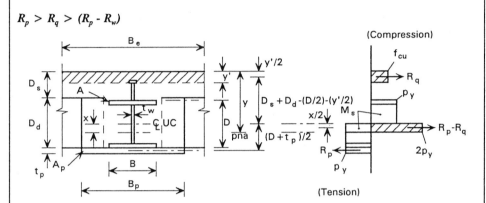

Moments about Centre Line (UC)

$$x = \frac{R_p - R_q}{2 p_y t_w}, \qquad y' = \frac{R_q}{0.45 f_{cu} B_e}$$

$$\boxed{M_c = M_s + R_q \left(D_d + D_s - \frac{D}{2} - \frac{R_q}{0.9 f_{cu} B_e} \right) + \frac{R_q}{2} (D + t_p) - \frac{(R_p - R_q)^2}{4 p_y t_w}}$$

Max value y'

$$y' = D_s \text{ when } D_d \geq D$$

$$y' = (D_s + D_d - D + T) \text{ when } D > D_d$$

		Job No:		Page *12* of *12*	Rev
	The Steel Construction Institute	Job Title	**Appendix B1.1 - Plastic Moment Capacity**		
Silwood Park, Ascot, Berks SL5 7QN Telephone: (01344) 23345 Fax: (01344) 22944		Subject	**Composite Section Based on Partial Shear Connection**		
CALCULATION SHEET		Client	Made by **DLM**	Date	
			Checked by **JWR**	Date	

CASE 4 PS (Partial Shear): pna in bottom UC flange

$$(R_q + R_s) > R_p > (R_q + R_w)$$

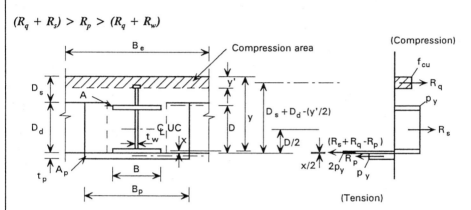

Moments taken about top of bottom flange plate

$$x = \frac{R_s + R_q - R_p}{2Bp_y}, \qquad y' = \frac{R_q}{0.45 f_{cu} B_e}$$

$$M_c = R_s D/2 + R_p t_p/2 + R_q (D_d + D_s - y'/2) - (R_s + R_q - R_p) x/2$$

$$\boxed{M_c = \frac{1}{2}(R_s D + R_p t_p) + R_q \left(D_d + D_s - \frac{R_q}{0.9 f_{cu} B_e} \right) - \frac{(R_s + R_p - R_q)^2}{4Bp_y}}$$

Max value y'

$$y' = D_s \text{ when } D_d \geq D$$

$$y' = (D_s + D_d - D + T) \text{ when } D > D_d$$

B1   Derivation of formulae

B1.2   Derivation of design formulae for Asymmetric Slimflor Beam (ASB) sections

The Steel Construction Institute	Job No:		Page *1* of *9*	Rev
	Job Title	*Appendix B1.2*		
	Subject	*Derivation of Design formulae for Asymmetric Sections*		
Silwood Park, Ascot, Berks SL5 7QN Telephone: (01344) 23345 Fax: (01344) 22944				
	Client		Made by	*RML*	Date
CALCULATION SHEET			Checked by	*JWR*	Date

B1.1   Plastic moment resistance of asymmetric steel section

The following parameters are used to determine the resistance of the ASB section:

B_b is the width of the bottom flange of the ASB section

B_t is the width of the top flange of the ASB section

D is the overall depth of the ASB section

d is the depth of the web $(= D - T_t - T_b)$

T_b is the thickness of the bottom flange

T_t is the thickness of the top flange

t_w is the thickness of the web

p_y is the design strength of steel

R_b is the tensile resistance of the bottom flange

R_t is the tensile resistance of the top flange

R_w is the tensile resistance of the web of depth, d

Depth of the plastic neutral axis, y_p, from top of the top flange:

The plastic neutral axis lies in the web when $(R_t - R_w) \le R_b \le (R_t + R_w)$, which is the case for all ASB sections, as shown in Figure B1.1. Hence,

$$y_p = \frac{d}{2} + T_t + \frac{(R_b - R_t)d}{2R_w} \tag{1}$$

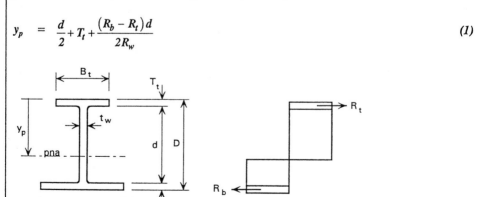

Figure B1.1   Steel section dimensions and plastic stress blocks

	Job No:		Page **2** of **9**	Rev
The Steel Construction Institute	Job Title	*Appendix B1.2*		
	Subject	*Derivation of Design formulae for Asymmetric Sections*		
Silwood Park, Ascot, Berks SL5 7QN Telephone: (01344) 23345 Fax: (01344) 22944	Client		Made by **RML**	Date
CALCULATION SHEET			Checked by **JWR**	Date

The plastic moment resistance of the steel section about its plastic neutral axis is given by:

$$M_s = R_t (y_p - 0.5 T_t) + R_b (D - y_p - 0.5 T_b)$$
$$+ R_w d/4 + (R_b - R_t)^2 d/(4 R_w) \qquad (2)$$

These formulae neglect the root radius of the steel section. Accurate section properties are given in Table 3.3.

R_b *may also be reduced due to transverse bending effects in the bottom flange (see Section 3.6.6).*

B1.2   Plastic moment resistance of composite section

The following additional parameters are used to determine the resistance of the composite ASB section:

B_e *is the effective breadth of the slab (= L/8)*
D_c *is the depth of concrete above the top of the ASB section*
D_d *is the depth of the steel decking*
D_s *is the depth of the solid slab above the decking*
f_{sb} *is the design shear-bond strength (= 0.6 N/mm²)*
f_{cu} *is the cube strength of concrete*
F_{sb} *is the compressive force transferred by longitudinal shear-bond*
R_c *is the compressive resistance of the slab of depth, D_s*
y_c *is the plastic neutral axis depth from the top of the slab*
y_{cc} *is the depth of concrete in compression*
L *is the beam span*

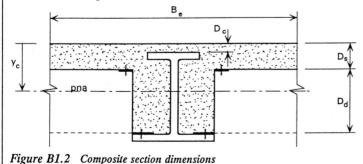

Figure B1.2 Composite section dimensions

	Job No:			Page *3* of *9*	Rev
The Steel Construction Institute	Job Title	*Appendix B1.2*			
	Subject	*Derivation of Design formulae for Asymmetric Sections*			
Silwood Park, Ascot, Berks SL5 7QN Telephone: (01344) 23345 Fax: (01344) 22944	Client		Made by	*RML*	Date
CALCULATION SHEET			Checked by	*JWR*	Date

The plastic moment resistance of the composite ASB section, as shown above in Figure B1.2, is established using the following formulae, depending on whether full or partial shear connection exists. The key parameters are:

Maximum compressive force in the slab due to longitudinal shear-bond transfer

$$F_{sb} = (B_t + T_t + d - 0.5\,t_w)\,\frac{L}{2}f_{sb} \tag{3}$$

Compressive resistance of the concrete slab

$$R_c = 0.45\,f_{cu}\,D_s\,B_e \tag{4}$$

where B_e = $L/8$

B1.2.1 Full shear connection

The plastic neutral axis of the section is first obtained by comparing the resistances of the concrete and steel components. Five distinct design cases exist, although only two or three of them apply in practical design cases using ASB sections. Case 4 is the most common case in practice. These cases are:

Case 1: Plastic neutral axis in the solid slab above the steel section.

This Case does **not** occur in practice, and is ignored in these derivations.

Case 2: Plastic neutral axis in the top flange of steel section. This Case is shown in Figure B1.3 and occurs when:

$$(R_w + R_b + R_t) \ge R_c \frac{D_c}{D_s} \ge (R_w + R_b - R_t) \tag{5}$$

It may be assumed that $y_c \approx D_c + 0.5\,T_t$

Full shear connection exists when $F_{sb} \ge R_c\,y_c/D_s$

	Job No:		Page **4** of **9**	Rev
![SCI logo] **The Steel Construction Institute** Silwood Park, Ascot, Berks SL5 7QN Telephone: (01344) 23345 Fax: (01344) 22944	Job Title	*Appendix B1.2*		
	Subject	*Derivation of Design formulae for Asymmetric Sections*		
	Client	Made by	*RML*	Date
CALCULATION SHEET		Checked by	*JWR*	Date

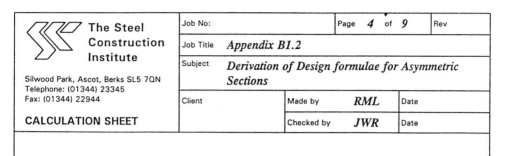

Figure B1.3   Case 2: Effective section and stress blocks

Plastic moment resistance about the centre of the top flange:

$$M_c = R_b\,(D - 0.5(T_b + T_t)) + \frac{R_w}{2}\,(d + T_t) + \frac{R_c}{2}\left(\frac{D_c^{\,2}}{D_s} + T_t\right) \tag{6}$$

This equation ignores the small moment due to the top flange about its mid-height.

R_b *may also be reduced due to transverse bending effects in the bottom flange.*

Case 3: Plastic neutral axis in the steel web and in the solid concrete slab

This case is shown in Figure B1.4, and

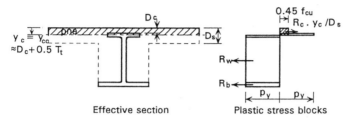

such that $D_c \le y_c \le D_s$ *and* $y_{cc} = y_c$

Full shear connection exists when $F_{sb} \ge R_c\,y_c/D_s$

		Job No:		Page 5 of 9	Rev

The Steel Construction Institute

Job Title *Appendix B1.2*

Subject *Derivation of Design formulae for Asymmetric Sections*

Silwood Park, Ascot, Berks SL5 7QN
Telephone: (01344) 23345
Fax: (01344) 22944

CALCULATION SHEET

Client		Made by	*RML*	Date
		Checked by	*JWR*	Date

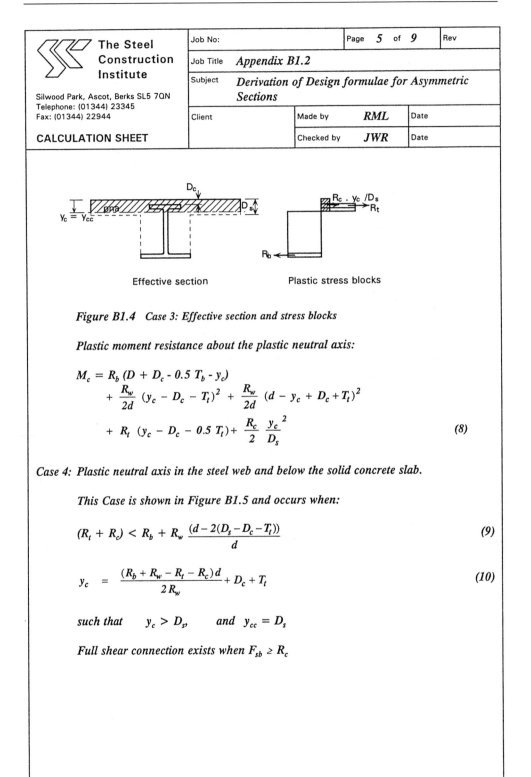

Effective section

Plastic stress blocks

Figure B1.4   Case 3: Effective section and stress blocks

Plastic moment resistance about the plastic neutral axis:

$$M_c = R_b \, (D + D_c - 0.5 \, T_b - y_c)$$
$$+ \frac{R_w}{2d} \, (y_c - D_c - T_t)^2 + \frac{R_w}{2d} \, (d - y_c + D_c + T_t)^2$$
$$+ R_t \, (y_c - D_c - 0.5 \, T_t) + \frac{R_c}{2} \, \frac{y_c^2}{D_s} \qquad (8)$$

Case 4: Plastic neutral axis in the steel web and below the solid concrete slab.

This Case is shown in Figure B1.5 and occurs when:

$$(R_t + R_c) < R_b + R_w \, \frac{(d - 2(D_s - D_c - T_t))}{d} \qquad (9)$$

$$y_c = \frac{(R_b + R_w - R_t - R_c) \, d}{2 \, R_w} + D_c + T_t \qquad (10)$$

such that $y_c > D_s$,   and $y_{cc} = D_s$

Full shear connection exists when $F_{sb} \geq R_c$

The Steel Construction Institute	Job No:			Page **6** of **9**		Rev
	Job Title	*Appendix B1.2*				
Silwood Park, Ascot, Berks SL5 7QN Telephone: (01344) 23345	Subject	*Derivation of Design formulae for Asymmetric Sections*				
Fax: (01344) 22944	Client		Made by	*RML*	Date	
CALCULATION SHEET			Checked by	*JWR*	Date	

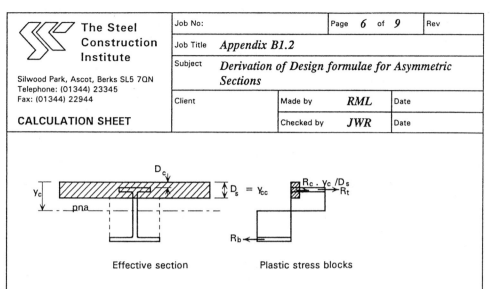

Effective section   Plastic stress blocks

Figure B1.5   Case 4: Effective section and stress blocks

The plastic moment resistance is obtained from Equation (8) using this value of y_c and replacing the final term $(0.5\ R_c\ y_c^2/D_s)$ by $R_c\ (y_c - 0.5\ D_s)$.

The Case is presented in the Equation for M_c shown in Section 3.8.9 in terms of the depth of the web in compression, y_s, rather than y_o which leads to a simplification of Equation (8) on sheet 5 of 9.

Case 5: Plastic neutral axis in the bottom flange.

This case does <u>not</u> occur in practice, and is ignored in these derivations.

B1.2.2 Partial shear connection

Partial shear connection exists in the above cases when:

Cases 2 and 3; $F_{sb} < R_c \dfrac{y_c}{D_s}$ 　　　　　　　　　(11)

Case 4; $F_{sb} < R_c$ 　　　　　　　　　(12)

The moment resistance of the composite section can be re-calculated for each Case of partial shear connection, as follows:

Case 6: Plastic neutral axis in the top flange of the steel section

This Case is shown in Figure B1.6, and occurs when partial shear connection exists in Case 2 and:

$F_{sb} \geq (R_b + R_w - R_t)$ 　　　　　　　　　(13)

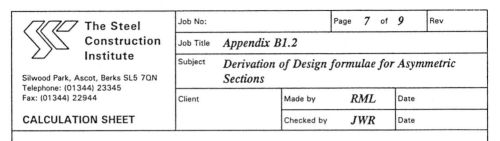

The Steel Construction Institute	Job No:		Page **7** of **9**	Rev
	Job Title *Appendix B1.2*			
Silwood Park, Ascot, Berks SL5 7QN Telephone: (01344) 23345 Fax: (01344) 22944	Subject *Derivation of Design formulae for Asymmetric Sections*			
	Client	Made by *RML*		Date
CALCULATION SHEET		Checked by *JWR*		Date

As an approximation, it may be assumed that F_{sb} acts at the centre of the solid slab above the steel section.

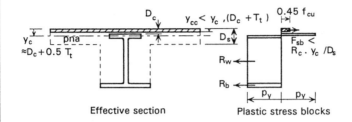

Figure B1.6   Case 6: Effective section and stress blocks for partial shear connection

The moment resistance about the centre of the top flange is:

$$M_c = R_b(D - 0.5(T_t + T_b)) + \frac{R_w}{2}(d + T_t) + \frac{F_{sb}}{2}(D_c + T_t) \qquad (14)$$

Case 7: Plastic neutral axis in the steel web below the depth of concrete in compression

This Case is shown in Figure B1.7, and occurs when partial shear connection exists in Cases 3 or 4, and:

$$F_{sb} < (R_b + R_w - R_t) \qquad (15)$$

The plastic neutral axis depth, y_c, from the top of the slab is obtained by replacing R_c by F_{sb} in Equation (10).

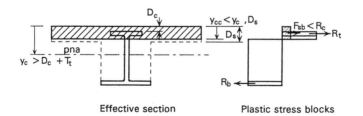

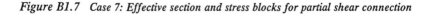

Effective section   Plastic stress blocks

Figure B1.7   Case 7: Effective section and stress blocks for partial shear connection

	Job No:		Page **8** of **9**	Rev
The Steel Construction Institute Silwood Park, Ascot, Berks SL5 7QN Telephone: (01344) 23345 Fax: (01344) 22944	Job Title	*Appendix B1.2*		
	Subject	*Derivation of Design formulae for Asymmetric Sections*		
CALCULATION SHEET	Client		Made by **RML**	Date
			Checked by **JWR**	Date

The moment resistance is obtained from Equation (8) using this value of y_c and replacing the final term $(0.5\ R_c\ y_c^2/D_s)$ by $F_{sb}\left(y_c - 0.5\dfrac{F_{sb}}{R_c}\ D_c\right)$.

B1.3   *Second moment of area of asymmetric steel section. Use the same parameters as in Appendix B1.1, and in addition:*

$$A_b = B_b\ T_b$$
$$A_t = B_t\ T_t$$
$$A_w = d\ t_w$$

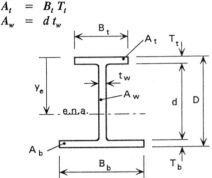

Figure B1.8   Case 8: Section dimensions for elastic analysis

Depth of elastic neutral axis from the top of the top flange, as shown in Figure B1.8, is:

$$y_e = \frac{A_t\ (0.5\ T_t) + A_b\ (D - 0.5\ T_b) + A_w\ 0.5(d + 2T_t)}{(A_t + A_b + A_w)} \qquad (16)$$

Second moment of area of the steel section:

$$I_x = A_t\ (y_e - 0.5\ T_t)^2 + A_b\ (D - y_e - 0.5\ T_b)^2$$

$$\qquad + A_w\ (y_e - 0.5\ (d + 2\ T_t))^2 + A_w\ d^2/12 \qquad (17)$$

B1.4   *Second moment of area of composite section*

Use same parameters as in Appendices B1.1 to B1.3 but add:

$$A_c = B_e\ D_s$$
$$A_c^1 = B_b\ D_d$$
$$\alpha_e = modular\ ratio\ of\ steel\ to\ concrete\ (\ = E_s/E_c)$$

The Steel Construction Institute	Job No:		Page **9** of **9**	Rev
	Job Title	*Appendix B1.2*		
Silwood Park, Ascot, Berks SL5 7QN Telephone: (01344) 23345 Fax: (01344) 22944	Subject	*Derivation of Design formulae for Asymmetric Sections*		
	Client		Made by **RML**	Date
CALCULATION SHEET			Checked by **JWR**	Date

where A_c^1 is the concrete encasement of depth D_d and width B_b.

Assume that the concrete section is uncracked for calculation of the elastic properties of the composite section, as shown in Figure B1.9.

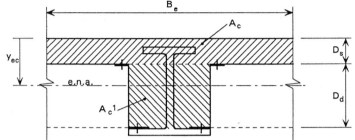

Figure B1.9   Composite section dimensions for elastic analysis

Depth of elastic neutral axis, y_{ec}, from the top of the steel section:

$$y_{ec} = \frac{A y_e + \frac{A_c^1}{\alpha_e}(D - 0.5D_d - T_b) + \frac{A_c}{\alpha_e}(0.5D_s - D_c)}{(A + (A_c + A_c^1)/\alpha_e)} \qquad (18)$$

where y_e is determined from Equation (16).

Second moment of area of the composite section

$$I_c = I_x + A(y_e - y_{ec})^2 + \frac{A_c}{\alpha_e}(0.5 D_s - D_c - y_{ec})^2 + \frac{A_c}{\alpha_e}\frac{D_s^2}{12}$$

$$+ \frac{A_c^1}{\alpha_e}(D - 0.5 D_d - T_b - y_{ec})^2 + \frac{A_c^1}{\alpha_e}\frac{D_d^2}{12} \qquad (19)$$

where I_x is determined from Equation (17),
and y_e is determined from Equation (16).
* A is the area of the steel section = $(A_b + A_t + A_w)$*
or A, y_e and I_x are obtained from the ASB section properties in Table 3.1.

B1   Derivation of formulae

B1.3   Formulae for plastic moment resistance of RHS Slimflor edge beam

- Non-composite
- Composite

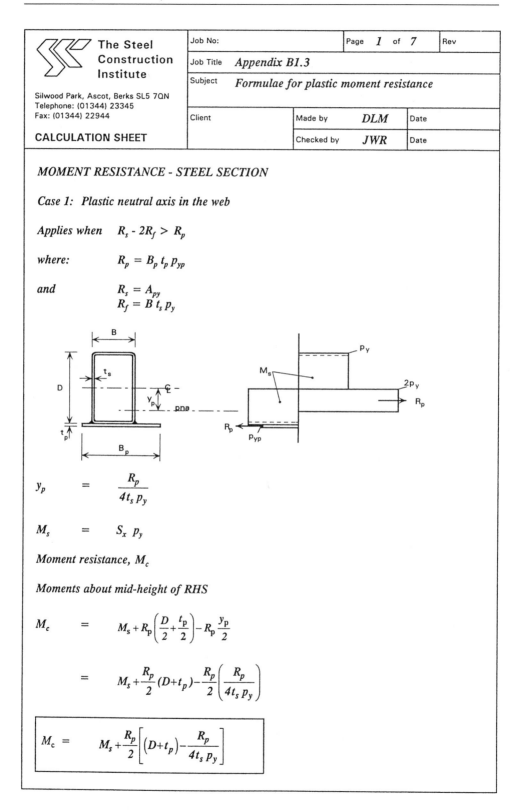

The Steel Construction Institute	Job No:		Page *1* of *7*	Rev
	Job Title	*Appendix B1.3*		
	Subject	*Formulae for plastic moment resistance*		
Silwood Park, Ascot, Berks SL5 7QN Telephone: (01344) 23345 Fax: (01344) 22944	Client		Made by	*DLM*	Date
CALCULATION SHEET			Checked by	*JWR*	Date

MOMENT RESISTANCE - STEEL SECTION

Case 1: Plastic neutral axis in the web

Applies when $R_s - 2R_f > R_p$

where: $R_p = B_p\, t_p\, p_{yp}$

and $R_s = A_{py}$
$R_f = B\, t_s\, p_y$

$$y_p = \frac{R_p}{4 t_s\, p_y}$$

$$M_s = S_x\, p_y$$

Moment resistance, M_c

Moments about mid-height of RHS

$$M_c = M_s + R_p\left(\frac{D}{2}+\frac{t_p}{2}\right) - R_p\frac{y_p}{2}$$

$$= M_s + \frac{R_p}{2}(D+t_p) - \frac{R_p}{2}\left(\frac{R_p}{4 t_s\, p_y}\right)$$

$$\boxed{M_c = M_s + \frac{R_p}{2}\left[(D+t_p) - \frac{R_p}{4 t_s\, p_y}\right]}$$

		Job No:		Page 2 of 7	Rev
The Steel Construction Institute		Job Title	*Appendix B1.3*		
		Subject	*Formulae for plastic moment resistance*		
Silwood Park, Ascot, Berks SL5 7QN Telephone: (01344) 23345 Fax: (01344) 22944		Client		Made by	DLM	Date
CALCULATION SHEET				Checked by	JWR	Date

Case 2: Plastic neutral axis in the bottom flange (RHS)

Applies when $R_s > R_p > R_s - 2R_f$

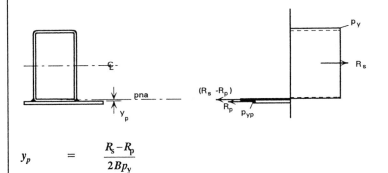

$$y_p \quad = \quad \frac{R_s - R_p}{2Bp_y}$$

Moment resistance, M_c

Moments about lower surface of flange plate

$$M_c \quad = \quad R_s \frac{D}{2} + R_p \frac{t_p}{2} - (R_s - R_p)\frac{y_p}{2}$$

$$\boxed{M_c \quad = \quad R_s \frac{D}{2} + R_p \frac{t_p}{2} - \frac{(R_s - R_p)^2}{4Bp_y}}$$

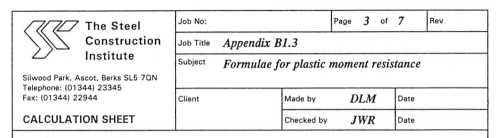

Case 3: Plastic neutral axis in the plate

Applies where $R_p > R_s$

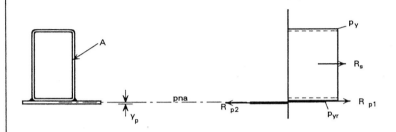

$$R_{p1} \quad = \quad B_p\,(t_p - y_p)\,p_{yp}$$

$$R_{p2} \quad = \quad B_p\,p_{yp}\,y_p$$

To find y_p, equate forces above and below pna

$$B_p\,p_{yp}\,y_p \quad = \quad R_s + (t_p - y_p)\,B_p\,p_{yp}$$

$$2\,B_p\,p_{yp}\,y_p \quad = \quad R_s + B_p\,p_{yp}\,t_p$$

$$\boxed{\therefore y_p \quad = \quad \frac{R_s + R_p}{2\,B_p\,p_{yp}}}$$

Moment resistance, M_c

Take moments about plastic neutral axis in plate

$$R_s\left(\frac{D}{2} + t_p - y_p\right) + \frac{R_{p1}}{2}(t_p - y_p) + R_{p2}\frac{y_p}{2} \qquad\qquad \text{but } R_{p2} = R_s + R_{p1}$$

$$= \quad R_s\left(\frac{D}{2} + t_p - y_p\right) + \frac{R_{p1}}{2}t_p + R_s\frac{y_p}{2} \qquad\qquad \text{but } R_{p1} = (t_p - y_p)\,B_p\,p_{yp}$$

$$= \quad R_s\left(\frac{D}{2} + t_p - y_p\right) + \frac{R_p}{2}(t_p - y_p) + R_s\frac{y_p}{2}$$

$$\boxed{\therefore M_c \quad = \quad R_s\left(\frac{D}{2} + t_p - \frac{y_p}{2}\right) + \frac{R_p}{2}(t_p - y_p)}$$

			Page **4** of **7**	Rev

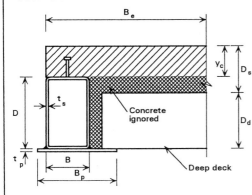

The Steel Construction Institute

Silwood Park, Ascot, Berks SL5 7QN
Telephone: (01344) 23345
Fax: (01344) 22944

CALCULATION SHEET

Job No:

Job Title   *Appendix B1.3*

Subject   *Formulae for plastic moment resistance*

Client

Made by   *DLM*   Date

Checked by   *JWR*   Date

COMPOSITE SECTION - FULL SHEAR CONNECTION *(General Notes)*

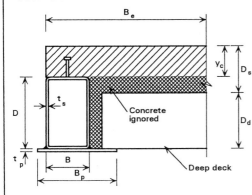
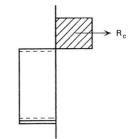

Effective breadth of concrete slab

$$B_e \quad = \quad \frac{L}{8} + \frac{B}{2}$$

y_c *is the maximum depth of concrete to be used for full shear connection.*

when $D_d > D,\ y_c = D_s$

Compression resistance of concrete slab

$$R_c \quad = \quad \left(\frac{L}{8} + \frac{B}{2}\right) y_c'\ 0.45\ f_{cu}$$

Tensile resistance of steel section $= \quad R_s + R_p$

Full shear connection exists when:

$$R_q \geq R_c \quad and \quad R_q \geq R_s + R_p$$

where: R_q $=$ *longitudinal shear force transfer by the shear connectors between points of zero and maximum moment*

Partial shear connection exists when:
$$R_q < R_c$$

The Steel Construction Institute Silwood Park, Ascot, Berks SL5 7QN Telephone: (01344) 23345 Fax: (01344) 22944 **CALCULATION SHEET**	Job No:		Page **5** of **7**	Rev
	Job Title *Appendix B1.3*		
	Subject *Formulae for plastic moment resistance*		
	Client	Made by **DLM**	Date
		Checked by **JWR**	Date

PARTIAL SHEAR CONNECTION

Case 1: Plastic neutral axis in top flange of RHS

Applies when $R_q < R_c$ *and* $R_s + R_p > R_q > R_s - 2R_f + R_p$

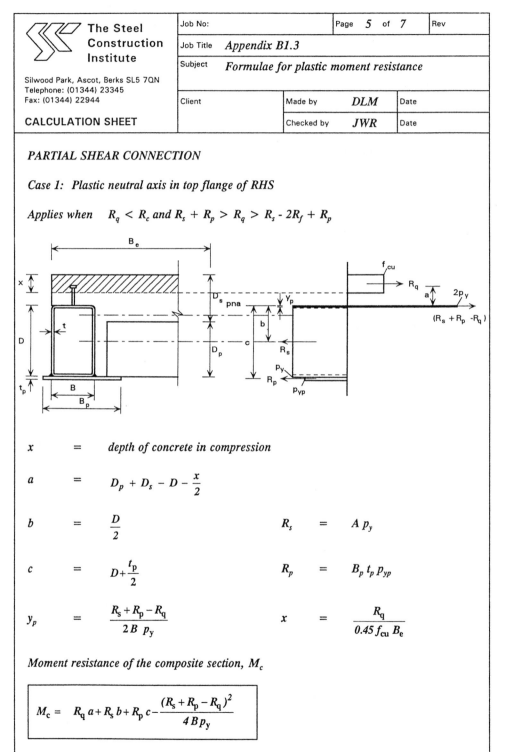

x $=$ depth of concrete in compression

a $=$ $D_p + D_s - D - \dfrac{x}{2}$

b $=$ $\dfrac{D}{2}$ $\qquad$ R_s $=$ $A\,p_y$

c $=$ $D + \dfrac{t_p}{2}$ $\qquad$ R_p $=$ $B_p\,t_p\,p_{yp}$

y_p $=$ $\dfrac{R_s + R_p - R_q}{2\,B\,p_y}$ $\qquad$ x $=$ $\dfrac{R_q}{0.45\,f_{cu}\,B_e}$

Moment resistance of the composite section, M_c

$$M_c = R_q\,a + R_s\,b + R_p\,c - \frac{(R_s + R_p - R_q)^2}{4\,B\,p_y}$$

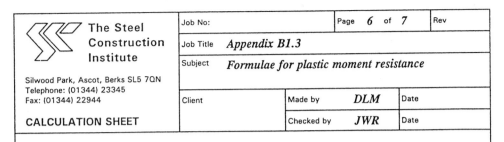

The Steel Construction Institute	Job No:			Page **6** of **7**		Rev
	Job Title	*Appendix B1.3*				
Silwood Park, Ascot, Berks SL5 7QN	Subject	*Formulae for plastic moment resistance*				
Telephone: (01344) 23345 Fax: (01344) 22944	Client			Made by	*DLM*	Date
CALCULATION SHEET				Checked by	*JWR*	Date

PARTIAL SHEAR CONNECTION

Case 2: *Plastic neutral axis in the web of the RHS*

Applies when $R_q < R_c$
$R_q + R_s > R_p + 2R_f$
$R_q < R_s - 2R_f + R_p$

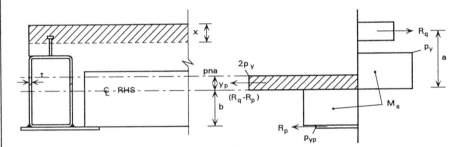

To find position y_p:

$$2 (2 p_y t_s y_p) = (R_q - R_p) \qquad \therefore \quad y_p = \frac{R_q - R_p}{4 p_y t_s}$$

$$a = D_s + D_p - \frac{D}{2} - \frac{x}{2} \qquad\qquad x = \frac{R_q}{0.45 f_{cu} B_e}$$

$$b = \frac{D}{2} + \frac{t_p}{2} \qquad\qquad M_s = S_x p_y \ (S_x \text{ for RHS})$$

Moment resistance of composite section, M_c

$$M_c = M_s + R_q a + R_p b - (R_q - R_p) \frac{y_p}{2}$$

$$\boxed{M_c = M_s + R_q a + R_p b - \frac{(R_q - R_p)^2}{8 p_y t_s}}$$

	Job No:		Page 7 of 7	Rev
The Steel Construction Institute	Job Title	*Appendix B1.3*		
Silwood Park, Ascot, Berks SL5 7QN Telephone: (01344) 23345 Fax: (01344) 22944	Subject	*Formulae for plastic moment resistance*		
	Client	Made by	*DLM*	Date
CALCULATION SHEET		Checked by	*JWR*	Date

PARTIAL SHEAR CONNECTION

Case 3: Plastic neutral axis in the lower flange of the RHS

Applies when $R_q < R_c$ and $R_q + R_s > R_p > R_q + R_s - 2R_f$

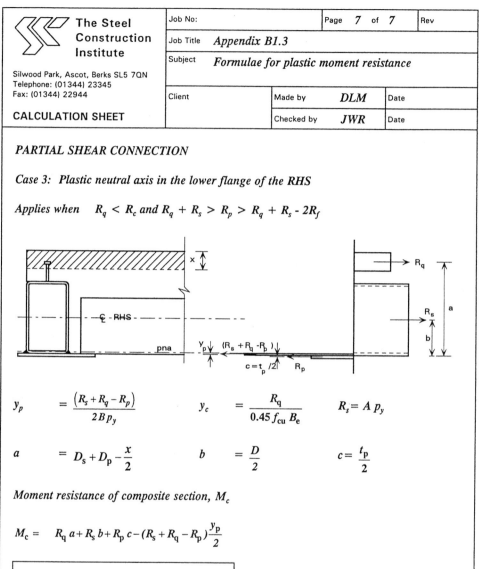

$$y_p = \frac{(R_s + R_q - R_p)}{2B p_y} \qquad y_c = \frac{R_q}{0.45 f_{cu} B_e} \qquad R_s = A p_y$$

$$a = D_s + D_p - \frac{x}{2} \qquad b = \frac{D}{2} \qquad c = \frac{t_p}{2}$$

Moment resistance of composite section, M_c

$$M_c = R_q a + R_s b + R_p c - (R_s + R_q - R_p)\frac{y_p}{2}$$

$$\boxed{M_c = R_q a + R_s b + R_p c - \frac{(R_s + R_q - R_p)^2}{4 B p_y}}$$

Note: *This case is unlikely to occur in practice.*

B2 Worked examples for Slimflor construction using deep profiled decking (SD225)

B2.1 Non-composite Slimflor beam (Type A)

		Job No.			Page		of		Rev.
The Steel Construction Institute						1		23	
		Job *Non Composite Slimflor Fabricated Beam*							
Silwood Park, Ascot, Berks SL5 7QN Telephone: (01344) 623345 Fax: (01344) 622944		Subject *Design Parameters*							
		Client			Made by *DLM*			Date	
CALCULATION SHEET					Checked by *JWR*			Date	

DESIGN OF NON-COMPOSITE BEAM

Beam and Decking are unpropped.

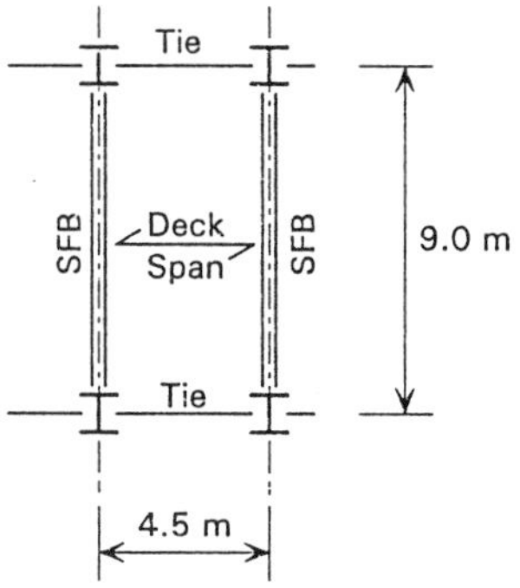

Section size: 254 × 254 × 132 UC
Grade S355
Lightweight concrete grade 30
LWC Density 19.0 kN/m³ (Wet)
18.0 kN/m³ (Dry)
To simplify the design use 19.0 kN/m³ throughout

PLAN (INTERNAL BAY)

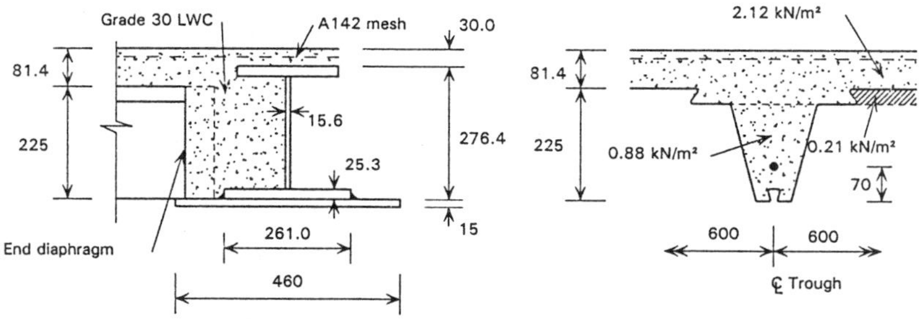

TYPICAL CROSS SECTION THROUGH SFB

Self Weight of Beam

$$132 \times 9.0 \times 9.81/10^3 \quad = \quad 11.7 \ kN$$

$$plate \ \frac{69}{168.9} \times 11.7 \quad = \quad \underline{4.8 \ kN}$$

$$16.5 \ kN$$

$$Weight/m^2 \quad = \quad \frac{16.5}{9 \times 4.5} \quad = \quad 0.4 \ kN/m^2$$

The Steel Construction Institute		Job No.		Page 2 of 23	Rev.
Silwood Park, Ascot, Berks SL5 7QN Telephone: (01344) 623345 Fax: (01344) 622944		Job *Non Composite Slimflor Fabricated Beam*			
		Subject *Loading*			
		Client	Made by *DLM*	Date	
CALCULATION SHEET			Checked by *JWR*	Date	

Concrete Weight kN/m^2

Slab (above trough)   2.12
Troughs   0.88
Crest Dovetail   -0.21
          Total   2.79

Concrete surrounding beam

$$\frac{0.46 \times 0.225 \times 19}{9.0 \times 4.5} = 0.035 \; say = \quad 0.05$$

          Total weight = $2.84 \; N/m^2$

Loading kN/m^2

Deck + end diaphragms   0.20
LW. concrete + reinforcement   2.84
Self weight of beam   0.4
          Total Dead: $3.44 \; kN/m^2$

Construction load   0.5
Services and finishes say,   0.25

Imposed kN/m^2

Occupancy   3.5
Partitions   1.0
          Total Imposed: $4.50 \; kN/m^2$

The 4.5 kN/m^2 value for imposed load reduces to 4.5 - 0.2 = 4.3 kN/m^2 based on a floor area of (9 × 4.5 m). See BS 6399 for further information.

Load Combinations

Case 1a   -   Out of balance loading in the construction stage.
                     Beam subjected to lateral torsional buckling (LTB).

Case 1b   -   Balanced loading - construction stage.
                     Beam subjected to LTB.

Case 2a   -   Imposed loading.
                     Beam assumed to be laterally restrained.

		Job No.		Page	3	of	23	Rev.

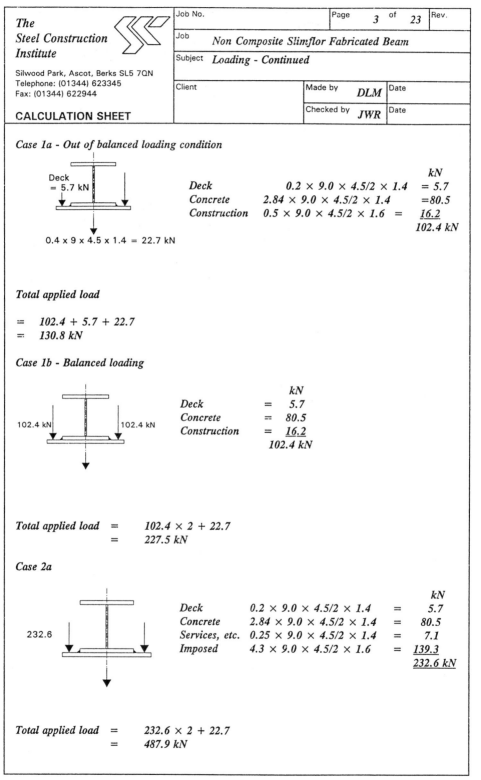

The
Steel Construction
Institute

Silwood Park, Ascot, Berks SL5 7QN
Telephone: (01344) 623345
Fax: (01344) 622944

CALCULATION SHEET

Job *Non Composite Slimflor Fabricated Beam*

Subject *Loading - Continued*

Client	Made by	*DLM*	Date
	Checked by	*JWR*	Date

Case 1a - Out of balanced loading condition

Deck
= 5.7 kN

0.4 x 9 x 4.5 x 1.4 = 22.7 kN

		kN
Deck	$0.2 \times 9.0 \times 4.5/2 \times 1.4$	= 5.7
Concrete	$2.84 \times 9.0 \times 4.5/2 \times 1.4$	=80.5
Construction	$0.5 \times 9.0 \times 4.5/2 \times 1.6$ =	16.2
		102.4 kN

Total applied load

= *102.4 + 5.7 + 22.7*
= *130.8 kN*

Case 1b - Balanced loading

102.4 kN   102.4 kN

		kN
Deck	=	5.7
Concrete	=	80.5
Construction	=	16.2
		102.4 kN

Total applied load = *102.4 × 2 + 22.7*
= *227.5 kN*

Case 2a

232.6

			kN
Deck	$0.2 \times 9.0 \times 4.5/2 \times 1.4$	=	5.7
Concrete	$2.84 \times 9.0 \times 4.5/2 \times 1.4$	=	80.5
Services, etc.	$0.25 \times 9.0 \times 4.5/2 \times 1.4$	=	7.1
Imposed	$4.3 \times 9.0 \times 4.5/2 \times 1.6$	=	139.3
			232.6 kN

Total applied load = *232.6 × 2 + 22.7*
= *487.9 kN*

The Steel Construction Institute	Job No.		Page 4 of 23	Rev.

	Job Non Composite Slimflor Fabricated Beam
	Subject *Transverse Plate Bending*

Silwood Park, Ascot, Berks SL5 7QN
Telephone: (01344) 623345
Fax: (01344) 622944

CALCULATION SHEET

Client	Made by	DLM	Date
	Checked by	JWR	Date

Transverse Plate Bending

Case 1b

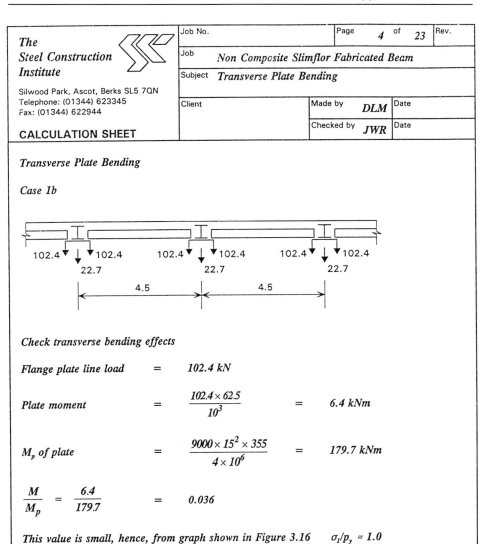

Check transverse bending effects

Flange plate line load $=$ *102.4 kN*

Plate moment $=$ $\dfrac{102.4 \times 62.5}{10^3}$ $=$ *6.4 kNm*

M_p *of plate* $=$ $\dfrac{9000 \times 15^2 \times 355}{4 \times 10^6}$ $=$ *179.7 kNm*

$\dfrac{M}{M_p}$ $=$ $\dfrac{6.4}{179.7}$ $=$ *0.036*

This value is small, hence, from graph shown in Figure 3.16 $\sigma_1/p_y \approx 1.0$

∴ neglect the influence of transverse bending of the bottom flange plate in the construction stage.

	Job No.		Page	5	of	23	Rev.

The
Steel Construction
Institute

Silwood Park, Ascot, Berks SL5 7QN
Telephone: (01344) 623345
Fax: (01344) 622944

CALCULATION SHEET

Job	*Non Composite Slimflor Fabricated Beam*
Subject	*Moment Capacity*

Client		Made by	*DLM*	Date
		Checked by	*JWR*	Date

Moment Capacity

Case 2 assumes pna in the UC flange ($R_s > R_p > R_w$)

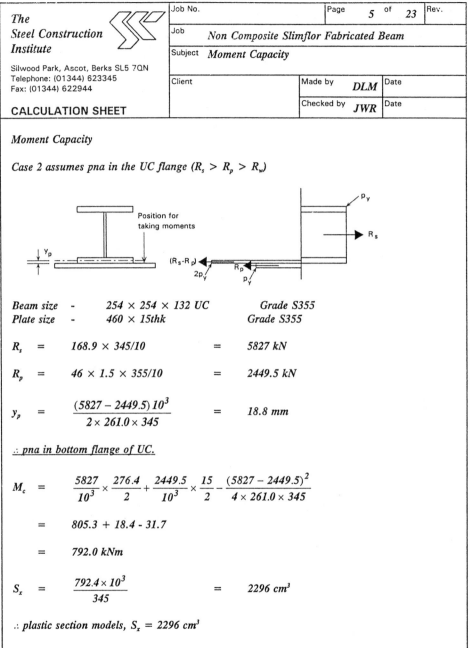

Beam size	-	*254 × 254 × 132 UC*	*Grade S355*
Plate size	-	*460 × 15thk*	*Grade S355*

R_s = $168.9 \times 345/10$ = $5827 \ kN$

R_p = $46 \times 1.5 \times 355/10$ = $2449.5 \ kN$

y_p = $\dfrac{(5827 - 2449.5) \, 10^3}{2 \times 261.0 \times 345}$ = $18.8 \ mm$

∴ pna in bottom flange of UC.

M_c = $\dfrac{5827}{10^3} \times \dfrac{276.4}{2} + \dfrac{2449.5}{10^3} \times \dfrac{15}{2} - \dfrac{(5827 - 2449.5)^2}{4 \times 261.0 \times 345}$

= $805.3 + 18.4 - 31.7$

= $792.0 \ kNm$

S_x = $\dfrac{792.4 \times 10^3}{345}$ = $2296 \ cm^3$

∴ plastic section models, $S_x = 2296 \ cm^3$

	Job No.			Page		of		Rev.
The *Steel Construction Institute*					6		23	
	Job	*Non Composite Slimflor Fabricated Beam*						
Silwood Park, Ascot, Berks SL5 7QN	Subject	*Elastic Section Properties*						
Telephone: (01344) 623345 Fax: (01344) 622944	Client			Made by	***DLM***		Date	
CALCULATION SHEET				Checked by	***JWR***		Date	

Case 1a

Out-of-balance loads in the construction stage. Beam assumed to be laterally unrestrained.

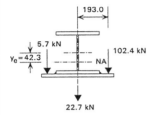

Net out-of-balance load   =   102.4 - 5.7   =   96.7 kN

Out-of-balance moment   =   96.7 × 193/10³   =   18.7 kNm

Elastic Section Properties

$A_{(UC)}$ =   168.9 cm²

A_p =   46 × 1.5   =   69.0 cm²

A_T =   168.9 + 69   =   237.9 cm²

y_e = $\dfrac{(276.4 + 15)\,69}{2 \times 237.9}$ =   42.3 mm

I_{xx} = $I_{x\,(UC)} + A\,y_e^2 + A_p \left(\dfrac{D}{2} + \dfrac{t_p}{2} - y_e \right)^2$

= $22575 + 168.9 \times 4.23^2 + 69 \left(\dfrac{276.4}{20} + \dfrac{15}{20} - 4.23 \right)^2$

=   25597 + 7377

=   32974 cm⁴

	Job No.		Page 7 of 23	Rev.
The **Steel Construction Institute**	Job	*Non Composite Slimflor Fabricated Beam*		
	Subject	*Buckling Parameter*		
Silwood Park, Ascot, Berks SL5 7QN				
Telephone: (01344) 623345	Client		Made by	*DLM*	Date
Fax: (01344) 622944					
CALCULATION SHEET			Checked by	*JWR*	Date

$$I_{yy} = I_{y\,(UC)} + \frac{t_p\, B_p^3}{12}$$

$$= 7519 + \frac{1.5 \times 46^3}{12}$$

$$= 19686 \; cm^4$$

$$r_{yy} = \left(\frac{19686}{237.9}\right)^{1/2} = 9.1 \; cm$$

$$I_{cf} = \frac{TB^3}{12} = \frac{2.53 \times 26.1^3}{12} = 3749 \; cm^4$$

$$I_{tf} = \frac{TB^3}{12} + \frac{t_p\, B_p^3}{12} = 3749 + \frac{1.5 \times 46^3}{12} = 15916 \; cm^4$$

Buckling Parameter, u

$$u = \left(\frac{4\, S_x^2\, \gamma}{A_T^2\, h_s^2}\right)^{1/4}$$

$$S_x = 2296 \; cm^3$$

$$\gamma = 1 - I_y/I_x = 1 - 19686/32974 = 0.403$$

$$u = \left(\frac{4 \times 2296^2 \times 0.403}{237.9^2 \times 26.38^2}\right)^{1/4} = 0.682$$

$$v = \left\{ \left[4N(1-N) + \frac{1}{20}(\lambda/x)^2 + \psi^2 \right]^{1/2} + \psi \right\}^{-1/2}$$

$$N = \frac{I_{cf}}{I_{cf} + I_{tf}} = \frac{3749}{3749 + 15916} = 0.191$$

$$N < 0.5, \quad \psi = 1\,(2N - 1) = 2 \times 0.191 - 1 = -0.618$$

Job No.				Page	8	of	23	Rev.

The Steel Construction Institute

Silwood Park, Ascot, Berks SL5 7QN
Telephone: (01344) 623345
Fax: (01344) 622944

CALCULATION SHEET

Job	Non Composite Slimflor Fabricated Beam
Subject	*Torsional Index*

Client		Made by	DLM	Date
		Checked by	JWR	Date

Torsional Index, x

$$x = 0.566\, h_s\, (A_T/J)^{1/2}$$

$$J = J_{(UC)} + \frac{1}{3}\, (t_p^3\, B_p)$$

$$= 319 + \frac{1}{3}\, (1.5^3 \times 46)$$

$$= 370.8\ cm^4$$

$$\therefore x = 0.566 \times 26.38\, (237.9/370.8)^{1/2}$$

$$= 11.96$$

$$\lambda = \frac{L_E}{r_y} = \frac{9000 \times 1.0}{91} = 98.9$$

$$\frac{\lambda}{x} = \frac{98.9}{11.96} = 8.269$$

$$v = \left\{ \left[(4 \times 0.191)(1 - 0.191) + \frac{8.269^2}{20} + 0.618^2 \right]^{1/2} - 0.618 \right\}^{-1/2}$$

$$= (1.484)^{-1/2}$$

$$= 0.821$$

<table>
<tr><td rowspan="3">**The
Steel Construction
Institute**

Silwood Park, Ascot, Berks SL5 7QN
Telephone: (01344) 623345
Fax: (01344) 622944

CALCULATION SHEET</td><td colspan="2">Job No.</td><td>Page **9** of **23**</td><td>Rev.</td></tr>
<tr><td colspan="4">Job *Non Composite Slimflor Fabricated Beam*</td></tr>
<tr><td colspan="4">Subject *Moment Capacity (M_b)*</td></tr>
<tr><td>Client</td><td colspan="2">Made by *DLM*</td><td>Date</td></tr>
<tr><td></td><td colspan="2">Checked by *JWR*</td><td>Date</td></tr>
</table>

$\therefore \lambda_{LT} = 1.0 \times 0.682 \times 0.821 \times 98.9$

$= 55.4$

From Table 11 (BS 5950: Part 1)

$p_b = 269.0 \ N/mm^2 \ (using \ p_y = 345 \ N/mm^2)$

$\therefore M_b = 2296 \times \dfrac{269.0}{10^3} = 617.6 \ kNm$

Case 2c

Out-of-balance loads

Use the rigorous method of analysis for torsion

$a = \left(\dfrac{EH}{GJ}\right)^{1/2}$ *where E/G is taken as 2.6*

$h \approx D - T/2 = 263.8 \ mm$

$H = \dfrac{h^2 I_{cf} I_{tf}}{I_y} = \dfrac{26.38^2 \times 3749 \times 15916}{19686} = 2.11 \times 10^6 \ cm^6$

$\therefore a = \left(\dfrac{2.6 \times 2.11 \times 10^6}{370.8}\right)^{1/2} = 121.6 \ cm$

$\dfrac{L}{a} = \dfrac{9000}{1216} = 7.4$

Using graph 1:

$\dfrac{\phi GJ}{T_q a} \approx 0.79 \ (see \ Sheet \ 12)$

$T_q = \underline{18.7 \ kNm}$

	Job No.		Page		of		Rev.
The Steel Construction Institute				10		23	
	Job *Non Composite Slimflor Fabricated Beam*						
Silwood Park, Ascot, Berks SL5 7QN	Subject *Buckling Check*						
Telephone: (01344) 623345 Fax: (01344) 622944	Client			Made by *DLM*		Date	
CALCULATION SHEET				Checked by *JWR*		Date	

$$\phi_{ULS} = \frac{0.53\,T_q\,a}{G\,J}$$

$$= \frac{0.79 \times 18.7 \times 10^6 \times 1216}{79000 \times 370.8 \times 10^4} \qquad = \qquad 0.0613 \; Rads \; (3.5°)$$

$$M_x \qquad = \qquad (102.4 + 5.7 + 22.7)\,9/8 \qquad = \qquad 147.2 \; kNm$$

$$M_{yT} \qquad = \qquad 147.2 \times 0.0613 \qquad = \qquad 9.02 \; kNm$$

$$\sigma_{byT} \qquad = \qquad M_{yT}/Z_y \qquad Z_y \qquad = \qquad I_y/y \qquad = \qquad \frac{19686}{13} \qquad = \qquad 1514 \; cm^3$$

$$\sigma_{byT} \qquad = \qquad \frac{9.02 \times 10^3}{1514} \qquad = \qquad 6.0 \; N/mm^2$$

Buckling Check

$$\frac{M_x}{M_b} + \left(\frac{\sigma_{byT} + \sigma_w}{p_y} \right) \left(1 + \frac{0.5\,M_x}{M_b} \right) \le 1.0$$

$$\sigma_w \qquad = \qquad E\,W_{no}\,\phi''$$

Calculate y_o

$$y_o \qquad = \qquad \frac{h_b.I_{tf} - h_t\,I_{cf}}{I_b + I_t} \qquad = \qquad \frac{9.58 \times 15916 - 16.79 \times 3749}{15916 + 3749} \qquad = \qquad 4.56 \; cm$$

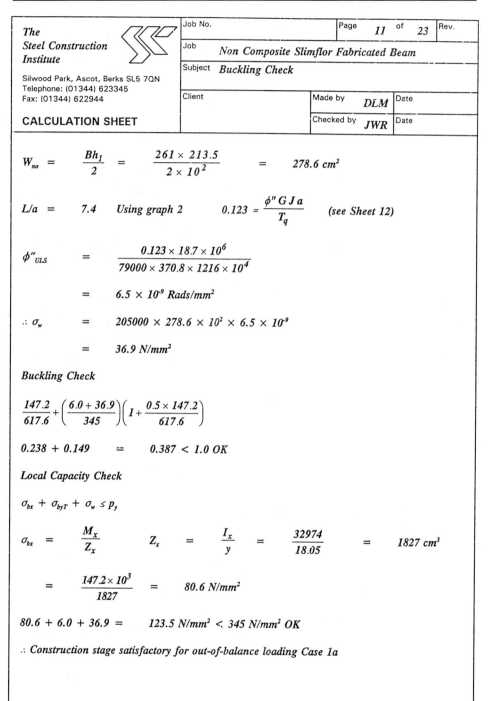

$$W_{no} = \frac{Bh_1}{2} = \frac{261 \times 213.5}{2 \times 10^2} = 278.6 \ cm^2$$

$$L/a = 7.4 \quad \text{Using graph 2} \quad 0.123 \approx \frac{\phi'' G J a}{T_q} \quad (see\ Sheet\ 12)$$

$$\phi''_{ULS} = \frac{0.123 \times 18.7 \times 10^6}{79000 \times 370.8 \times 1216 \times 10^4}$$

$$= 6.5 \times 10^{-9} \ Rads/mm^2$$

$$\therefore \sigma_w = 205000 \times 278.6 \times 10^2 \times 6.5 \times 10^{-9}$$

$$= 36.9 \ N/mm^2$$

Buckling Check

$$\frac{147.2}{617.6} + \left(\frac{6.0 + 36.9}{345}\right)\left(1 + \frac{0.5 \times 147.2}{617.6}\right)$$

$$0.238 + 0.149 = 0.387 < 1.0 \ OK$$

Local Capacity Check

$$\sigma_{bx} + \sigma_{byT} + \sigma_w \leq p_y$$

$$\sigma_{bx} = \frac{M_x}{Z_x} \qquad Z_x = \frac{I_x}{y} = \frac{32974}{18.05} = 1827 \ cm^3$$

$$= \frac{147.2 \times 10^3}{1827} = 80.6 \ N/mm^2$$

$$80.6 + 6.0 + 36.9 = 123.5 \ N/mm^2 < 345 \ N/mm^2 \ OK$$

$\therefore$ **Construction stage satisfactory for out-of-balance loading Case 1a**

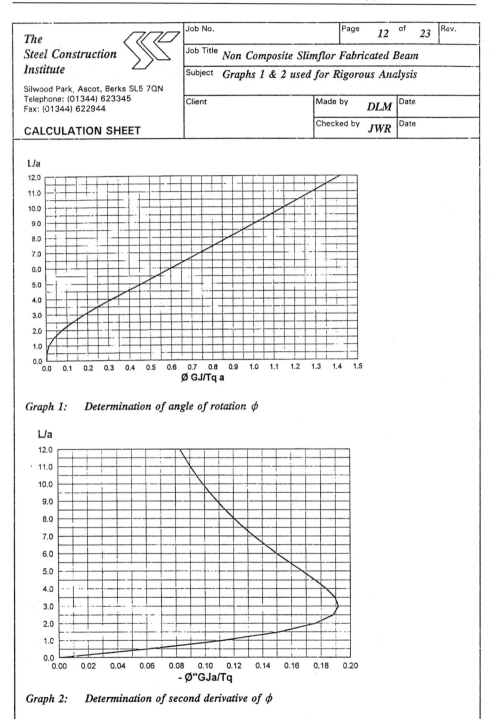

The		Job No.			Page *12* of *23*	Rev.
The						
Steel Construction		Job Title	*Non Composite Slimflor Fabricated Beam*			
Institute		Subject	*Graphs 1 & 2 used for Rigorous Analysis*			

Silwood Park, Ascot, Berks SL5 7QN
Telephone: (01344) 623345
Fax: (01344) 622944

CALCULATION SHEET

| Client | Made by | *DLM* | Date |
| | Checked by | *JWR* | Date |

Graph 1:   Determination of angle of rotation ϕ

Graph 2:   Determination of second derivative of ϕ

Note: *These two graphs have been modified from the graphs shown in Reference 9*

		Job No.				Page		of		Rev.
The *Steel Construction* *Institute*		Job					13		23	
			Non Composite Slimflor Fabricated Beam							
Silwood Park, Ascot, Berks SL5 7QN	Subject		*Flange Movement in the Construction Stage*							
Telephone: (01344) 623345 Fax: (01344) 622944		Client				Made by	*DLM*		Date	
CALCULATION SHEET						Checked by	*JWR*		Date	

Horizontal Deflection (Top Flange of Beam)

N.B. This is an optional check

L/a = *7.4*

$$\frac{\phi G J}{T_q a} \approx 0.79 \ (\text{Graph 1 on sheet 12})$$

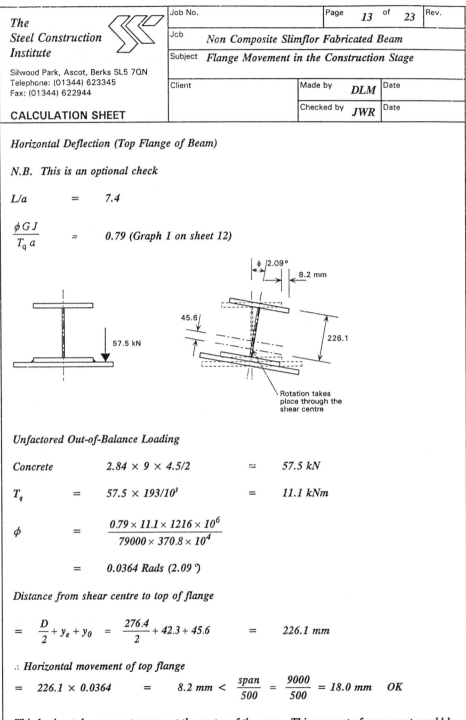

Rotation takes place through the shear centre

Unfactored Out-of-Balance Loading

Concrete $2.84 \times 9 \times 4.5/2$ =   *57.5 kN*

T_q = $57.5 \times 193/10^3$ =   *11.1 kNm*

$$\phi = \frac{0.79 \times 11.1 \times 1216 \times 10^6}{79000 \times 370.8 \times 10^4}$$

     =   *0.0364 Rads (2.09 °)*

Distance from shear centre to top of flange

$$= \frac{D}{2} + y_e + y_0 = \frac{276.4}{2} + 42.3 + 45.6 = 226.1 \ mm$$

∴ Horizontal movement of top flange

$$= \ 226.1 \times 0.0364 \quad = \quad 8.2 \ mm < \frac{span}{500} = \frac{9000}{500} = 18.0 \ mm \quad OK$$

This horizontal movement occurs at the centre of the span. This amount of movement would be considered satisfactory. The deck connected either side of the beam to the flange plate plus the end diaphragms will provide additional stiffness. This additional stiffness has not been included in the above.

The		Job No.		Page	14	of	23	Rev.
Steel Construction Institute		Job	*Non Composite Slimflor Fabricated Beam*					
		Subject	*Ultimate Limit State: Moment Capacity*					
Silwood Park, Ascot, Berks SL5 7QN Telephone: (01344) 623345 Fax: (01344) 622944		Client		Made by	*DLM*	Date		
CALCULATION SHEET				Checked by	*JWR*	Date		

Case 2a

Assume vertical shear applied to the bottom flange plate except for the self weight of beam. This is conservative as it ignores the shear transfer through the concrete.

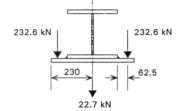

Total load   =   487.9 kN (see Sheet 3)

plate 15 mm thick < 16 mm

$$\therefore p_y = 355 \ N/mm^2$$

Check influence of transverse plate bending

$$Plate \ moment \quad = \quad 232.6 \times 62.5/10^3 \quad = \quad 14.5 \ kNm$$

$$M_p \ (plate) \quad = \quad \frac{9000 \times 15^2 \times 355}{4 \times 10^6} \quad = \quad 179.7 \ kNm$$

$$\frac{M}{M_p} \quad = \quad \frac{14.5}{179.7} \quad = \quad 0.08$$

From Figure 3.16 limit of $\sigma_1/p_y \approx 0.98$

$\therefore$ maximum value of $\sigma_1 = 0.98 \times 355 = 348 \ N/mm^2$

Hence the value of p_y in the bottom flange plate will be limited to 348 N/mm^2

Moment capacity

$$R_p \ = 46 \times 1.5 \times 348/10 = 2401.2 \ kN, \ R_s = 5827 \ kN$$

Assume pna in the UC flange plate ($R_s > R_p > R_w$)

$$\therefore y_p \ = \quad \frac{(5827 - 2401.2) \ 10^3}{2 \times 261.3 \times 345} \quad = \quad 19.02$$

$\therefore$ pna in bottom flange of UC

$$M_c = \frac{5827}{10^3} \times \frac{276.4}{2} + \frac{2401.2}{10^3} \times \frac{15}{2} - \frac{(5827 - 2401.2)^2}{4 \times 261 \times 345}$$

$$= 805.3 + 18.0 - 32.6$$

$$= 790.7 \ kNm$$

Maximum longitudinal moment $\quad = \quad 487.9 \times \dfrac{9.0}{8} \quad = \quad 548.9 \ kNm$

Since 548.9 kNm < 790.7 kNm, beam OK.

The Steel Construction Institute	Job No.		Page 16 of 23	Rev.
	Job Non Composite Slimflor Fabricated Beam			
Silwood Park, Ascot, Berks SL5 7QN Telephone: (01344) 623345 Fax: (01344) 622944	Subject Ultimate Limit State Vertical Shear			
CALCULATION SHEET	Client		Made by DLM	Date
			Checked by JWR	Date

Vertical Shear

Case 2a

F_v =   487.9 kN/2   =   244.0 kN

Shear Capacity

P_v = $0.6\,p_y\,A_v$ A_v = $t_w\,D$

= $\dfrac{0.6 \times 345 \times 15.6 \times 276.4}{10^3}$

=   892.6 kN > 244.0 kN OK

Note:

No need to check for combination of moment and high shear (> 0.6 P_v) as applied loads are uniformly distributed.

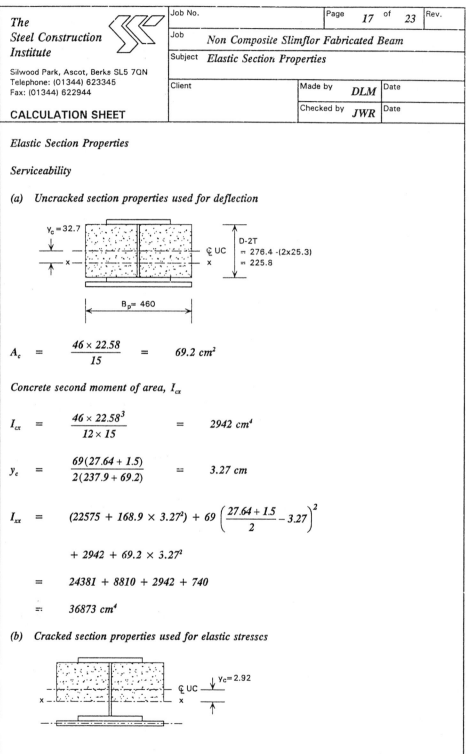

The
Steel Construction
Institute

Silwood Park, Ascot, Berks SL5 7QN
Telephone: (01344) 623345
Fax: (01344) 622944

CALCULATION SHEET

Job   Non Composite Slimflor Fabricated Beam

Subject   *Elastic Section Properties*

Client

Made by   *DLM*   Date

Checked by   *JWR*   Date

Elastic Section Properties

Serviceability

(a) *Uncracked section properties used for deflection*

$Y_c = 32.7$

$\mathcal{C}$ UC

D-2T
$= 276.4 - (2 \times 25.3)$
$= 225.8$

$B_p = 460$

$$A_c = \frac{46 \times 22.58}{15} = 69.2 \ cm^2$$

Concrete second moment of area, I_{cx}

$$I_{cx} = \frac{46 \times 22.58^3}{12 \times 15} = 2942 \ cm^4$$

$$y_c = \frac{69(27.64 + 1.5)}{2(237.9 + 69.2)} = 3.27 \ cm$$

$$I_{xx} = (22575 + 168.9 \times 3.27^2) + 69 \left(\frac{27.64 + 1.5}{2} - 3.27 \right)^2$$

$$+ 2942 + 69.2 \times 3.27^2$$

$$= 24381 + 8810 + 2942 + 740$$

$$= 36873 \ cm^4$$

(b) *Cracked section properties used for elastic stresses*

$y_c = 2.92$

$\mathcal{C}$ UC

The Steel Construction Institute	Job No. Page *18* of *23* Rev.
	Job *Non Composite Slimflor Fabricated Beam*
Silwood Park, Ascot, Berks SL5 7QN	Subject *Elastic Section Properties - Continued*
Telephone: (01344) 623345 Fax: (01344) 622944	Client Made by *DLM* Date
CALCULATION SHEET	Checked by *JWR* Date

$$k_1 = \frac{46}{2 \times 15} = 1.533$$

$$k_2 = 237.9 + \frac{46}{15}\left(\frac{22.58}{2}\right) = 272.5$$

$$k_3 = \frac{46}{2 \times 15}\left(\frac{22.58}{2}\right)^2 - 69\left(\frac{29.14}{2}\right) = -809.9$$

$$y_c = \frac{-272.5 + (272.5^2 - 4(1.533)(-809.9))^{1/2}}{2 \times 1.533}$$

$$= \frac{-272.5 + 281.5}{2 \times 1.533}$$

$$= 2.92 \ cm$$

$$A_c = \frac{46}{15}\left[\left(\frac{22.58}{2}\right) + 2.92\right] = 43.6 \ cm^2$$

$$I_{xx} = (22575 + 168.9 \times 2.92^2) + 69\left(\frac{27.64 + 1.5}{2} - 2.92\right)^2$$

$$+ \frac{46}{12 \times 15}\left[\left(\frac{22.58}{2}\right) + 2.92\right]^3 + 43.6\left[\left(\frac{22.58}{4}\right) - \frac{2.92}{2}\right]^2$$

$$= 24015 + 9365 + 733 + 764$$

$$= 34877 \ cm^4$$

$$Z(Concrete) = \frac{34877 \times 15}{\frac{22.58}{2} + 2.92} = 36816 \ cm^3$$

Z (Steel)

$$Top \ (Compression) = \frac{34877}{\frac{27.64}{2} + 2.92} = 2083 \ cm^3$$

$$Bottom \ (Tension) = \frac{34877}{\frac{27.64}{2} + 1.5 - 2.92} = 2813 \ cm^3$$

		Job No.		Page 19 of 23	Rev.

The
Steel Construction
Institute

Silwood Park, Ascot, Berks SL5 7QN
Telephone: (01344) 623345
Fax: (01344) 622944

CALCULATION SHEET

Job   *Non Composite Slimflor Fabricated Beam*

Subject   *Serviceability Stresses*

Client

Made by   *DLM*   Date

Checked by   *JWR*   Date

Elastic Stress

Loading	kN/m^2
Construction stage dead	*3.44 (steel beam only)*
Imposed	*4.3 - composite*
Service and Finishes	*0.25 - composite*

Construction Stage (steel beam only)

Loading

$$W \quad = \quad 3.44 \times 9.0 \times 4.5$$

$$= \quad 139.3 \ kN$$

$$M_x \quad = \quad 139.3 \times \frac{9.0}{8} \quad = \quad 156.7 \ kNm$$

$$I_{xx} \quad = \quad 32974 \ cm^4$$

$$Z \ (top) \quad = \quad \frac{32974}{18.05} \quad = \quad 1827 \ cm^3$$

$$Z \ (bottom) \quad = \quad \frac{32974}{11.09} \quad = \quad 2973 \ cm^3$$

Steel Stress

$$Compression \quad = \quad \frac{156.7 \times 10^3}{1827} \quad = \quad 85.7 \ N/mm^2$$

$$Tension \quad = \quad \frac{156.7 \times 10^3}{2973} \quad = \quad 52.7 \ N/mm^2$$

Composite Stage (Cracked Section)

$$W \quad = \quad (4.3 + 0.25) \ 9 \times 4.5 \quad = \quad 184.3 \ kN$$

$$M_x \quad = \quad 184.3 \times \frac{9.0}{8} \quad = \quad 207.3 \ kNm$$

The
Steel Construction
Institute

Silwood Park, Ascot, Berks SL5 7QN
Telephone: (01344) 623345
Fax: (01344) 622944

CALCULATION SHEET

Job: *Non Composite Slimflor Fabricated Beam*

Subject: *Serviceability Stresses - Continued*

Client | Made by **DLM** Date
Checked by **JWR** Date

Concrete Stress

$$\text{Compression} \quad = \quad \frac{207.3 \times 10^3}{36816} \quad = \quad 5.6\ N/mm^2 < \frac{30}{2} \quad = \quad 15\ N/mm^2\ OK$$

Steel Stress

$$\text{Compression} \quad = \quad \frac{207.3 \times 10^3}{2083} \quad = \quad 99.5\ N/mm^2$$

$$\text{Tension} \quad = \quad \frac{207.3 \times 10^3}{2813} \quad = \quad 73.7\ N/mm^2$$

Steel stress for construction stage + composite stages

$$\text{Compression } (\sigma_c) \quad = \quad 85.7 + 99.5 = 185.2\ N/mm^2 < p_y = 345\ N/mm^2\ \ OK$$

$$\text{Tension } (\sigma_t) \quad = \quad 52.7 + 73.7 = 126.4\ N/mm^2 < p_y = 355\ N/mm^2\ \ OK$$

The **Steel Construction Institute**	Job No. — Page *21* of *23* Rev.
	Job — *Non Composite Slimflor Fabricated Beam*
Silwood Park, Ascot, Berks SL5 7QN Telephone: (01344) 623345 Fax: (01344) 622944	Subject — *Transverse Plate Bending - Elastic Condition*
	Client — Made by *DLM* Date
CALCULATION SHEET	Checked by *JWR* Date

Transverse Plate Bending - Elastic Condition

Elastic condition

Longitudinal stress, σ_t

σ_t from Sheet 20   =   126.4 N/mm²

plate bending   kN/m²

Deck	=	0.20
Concrete	=	2.84

$3.04 \times 9 \times 4.5/2$ =   61.6 kN

Imposed	=	4.23
Services	=	0.25
		4.55

$$W = 4.55 \times \frac{9.0 \times 4.5}{2} = 92.1\ kN$$

∴ Flange plate line load = 61.6 + 92.1 kN
= 153.7 kN

plate bending = 153.7 × 62.5/10³ = 9.6 kNm

plate modulus, Z_e

$$Z_e = \frac{1000 \times 15^2}{6 \times 10^3} = 37.5\ cm^3\ per\ m$$

per span = 37.5 × 9.0 = 337.5 cm³

$$\sigma_{tt} = \frac{9.6 \times 10^6}{337.5 \times 10^3} = 28.4\ N/mm^2$$

Combined stress must be less than or equal to p_y

i.e. $(\sigma_{tt}^2 + \sigma_t^2 + \sigma_{tt}\ \sigma_t)^{\frac{1}{2}} \not> p_y$

$(28.4^2 + 126.4^2 + 28.4 \times 126.4)^{\frac{1}{2}}$

= 142.7 N/mm² < 355 N/mm²

Serviceability stresses for concrete and steel OK

		Job No.		Page 22 of 23	Rev.

The Steel Construction Institute

Silwood Park, Ascot, Berks SL5 7QN
Telephone: (01344) 623345
Fax: (01344) 622944

CALCULATION SHEET

Job	Non Composite Slimflor Fabricated Beam
Subject	*Deflection*

Client		Made by	DLM	Date
		Checked by	JWR	Date

Deflection

Construction stage (steel beam only)

Loading

Self weight	=	$0.4 \times 9.0 \times 4.5$	=	162 kN
Decking	=	$0.2 \times 9.0 \times 4.5$	=	8.1 kN
Concrete	=	$2.84 \times 9.0 \times 4.5$	=	115.0 kN
		W	=	139.3 kN

$$\delta_{cl} = \frac{5 \times 9000^3 \times 139.3}{384 \times 205 \times 32974 \times 10^4} = 19.6 \ mm$$

Imposed Loading

$$W = 4.3 \times 9.0 \times 4.5 = 174.0 \ kN$$

$$\delta_{cl} = \frac{5 \times 174 \times 9000^3}{384 \times 205 \times 36873 \times 10^4} = 21.9 \ mm \quad < \quad (L/360 = 25.0 \ mm) \ OK$$

Services

$$\delta_s = 21.9 \times 0.25/4.3 = 1.3 \ mm$$

$$\delta_{total} = 19.6 + 21.9 + 1.3 = 42.8 \ mm \qquad (L/210) \ OK$$

Note:

Where construction stage deflections of the steel deck are considered to be high, then account should be taken of the increased concrete weight due to ponding.

Refer to deck manufacturer's literature.

		Job No.		Page	23	of	23	Rev.

The
Steel Construction
Institute

Silwood Park, Ascot, Berks SL5 7QN
Telephone: (01344) 623345
Fax: (01344) 622944

CALCULATION SHEET

Job	*Non Composite Slimflor Fabricated Beam*
Subject	*Natural Frequency*

Client		Made by	*DLM*	Date	
		Checked by	*JWR*	Date	

Natural Frequency

Loading	$=$	*Dead + 10% Imposed (not including partitions)*
	$=$	*(3.44 + 0.25) + 10% of 3.3 kN/m²*
	$=$	*3.69 + 0.33*
	$=$	*4.02 kN/m²*

$$W \quad = \quad 4.02 \times 9.0 \times 4.5 \quad = \quad 162.8 \; kN$$

$$\delta_f \quad = \quad \frac{5 \times 162.8 \times 9000^3}{384 \times 205 \times 36873 \times 10^4} \quad = \quad 20.4 \; mm$$

$$Frequency \quad = \quad \frac{18}{\sqrt{20.4}}$$

$$= \quad 4.0 \; Hz \; OK$$

∴ *Floor structure satisfactory for natural frequency.*

If this check had not satisfied the 4.0 Hz limit, then reference (16), dealing with vibration, provides a more rigorous method of analysis.

B2 Worked examples for Slimflor construction using deep profiled decking (SD225)

B2.2 Composite Slimflor beam (Type B)

		Job No.		Page	1	of	22	Rev.

The Steel Construction Institute

Silwood Park, Ascot, Berks SL5 7QN
Telephone: (01344) 623345
Fax: (01344) 622944

CALCULATION SHEET

Job	*Composite Slimflor Fabricated Beam*
Subject	*Design Parameters*

Client		Made by	DLM	Date
		Checked by	JWR	Date

DESIGN OF COMPOSITE BEAM (TYPE B)

Unpropped construction

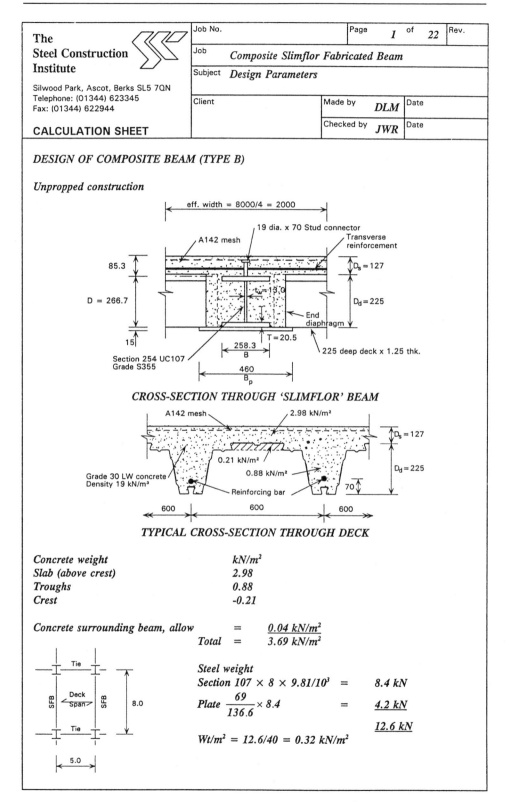

CROSS-SECTION THROUGH 'SLIMFLOR' BEAM

TYPICAL CROSS-SECTION THROUGH DECK

Concrete weight	kN/m^2
Slab (above crest)	*2.98*
Troughs	*0.88*
Crest	*-0.21*

Concrete surrounding beam, allow $\quad = \quad \underline{0.04\ kN/m^2}$

$\qquad\qquad Total \quad = \quad 3.69\ kN/m^2$

Steel weight

Section $107 \times 8 \times 9.81/10^3 \quad = \quad \underline{8.4\ kN}$

Plate $\dfrac{69}{136.6} \times 8.4 \qquad = \quad \underline{4.2\ kN}$

$\qquad\qquad\qquad\qquad\qquad\qquad \underline{12.6\ kN}$

$Wt/m^2 = 12.6/40 = 0.32\ kN/m^2$

		Job No.			Page 2 of 22	Rev.

The
Steel Construction
Institute

Silwood Park, Ascot, Berks SL5 7QN
Telephone: (01344) 623345
Fax: (01344) 622944

CALCULATION SHEET

Job   *Composite Slimflor Fabricated Beam*

Subject   *Assumptions*

Client

Made by   *DLM*   Date

Checked by   *JWR*   Date

General Assumptions

1)   *Laterally unrestrained in the construction stage.*

2)   *Beam assumed to be torsionally fixed and warping free at the supports for the construction stage.*

Case 1

a)   *Lateral torsional buckling (LTB) check has to be combined with torsion.*

b)   *Max loading condition for LTB during construction.*

Case 2

a)   *Beam is considered to be laterally restrained. Max factored moment applied to composite section.*

The main assumptions are:

(a)   *Unpropped simply supported beams are subject to uniformly distributed loading.*

(b)   *Use only plastic or compact cross-sections.*

(c)   *Plastic analysis of the cross-section is based on rectangular stress blocks.*

(d)   *Moments and forces are determined using factored loads.*

(e)   *Serviceability checks are determined using unfactored loads. To ensure that irreversible deformation (under normal service loads) does not occur in the steel, the extreme fibre stress is limited to p_y. The in situ concrete stress is likewise limited to $0.5 f_{cu}$.*

(f)   *Deflections of beam are limited to span/360 under imposed loads and span/200 under total load. Pre-cambering should be considered when the total deflection exceeds this limit. These limitations for deflection apply to buildings of general usage. In addition, it is a requirement of BS 5950: Part 1 that due allowance should be made where deflections under serviceability loads could impair the strength or efficiency of the structure or its components or cause damage to the finishings.*

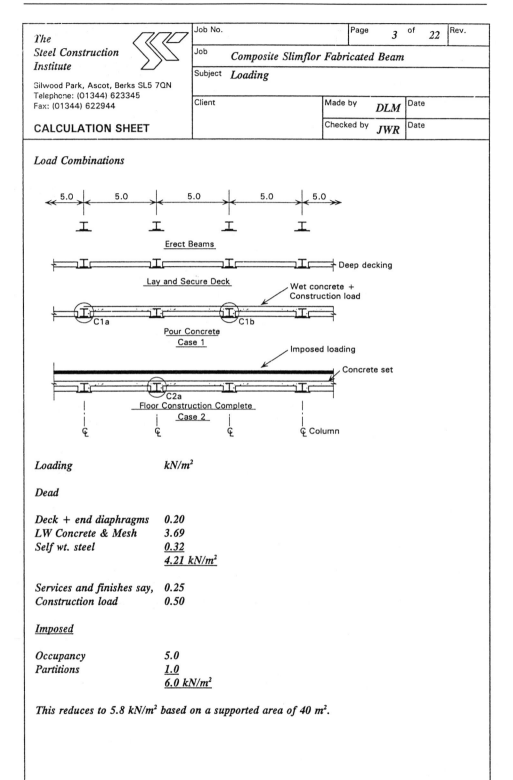

| | Job No. | | | Page | 3 | of | 22 | Rev. |

The
Steel Construction
Institute

Silwood Park, Ascot, Berks SL5 7QN
Telephone: (01344) 623345
Fax: (01344) 622944

CALCULATION SHEET

Job   *Composite Slimflor Fabricated Beam*

Subject   *Loading*

Client

Made by   *DLM*   Date

Checked by   *JWR*   Date

Load Combinations

Loading kN/m^2

Dead

Deck + end diaphragms	*0.20*
LW Concrete & Mesh	*3.69*
Self wt. steel	*0.32*
	4.21 kN/m²

Services and finishes say,	*0.25*
Construction load	*0.50*

Imposed

Occupancy	*5.0*
Partitions	*1.0*
	6.0 kN/m²

This reduces to 5.8 kN/m² based on a supported area of 40 m².

The Steel Construction Institute		Job No.		Page 4 of 22	Rev.
		Job Composite Slimflor Fabricated Beam			
		Subject Transverse Bending Effects			
Silwood Park, Ascot, Berks SL5 7QN Telephone: (01344) 623345 Fax: (01344) 622944		Client		Made by DLM	Date
CALCULATION SHEET				Checked by JWR	Date

Transverse Bending Effects (Deck Unpropped)

Bending to bottom flange plate

Construction Stage

Case 1b

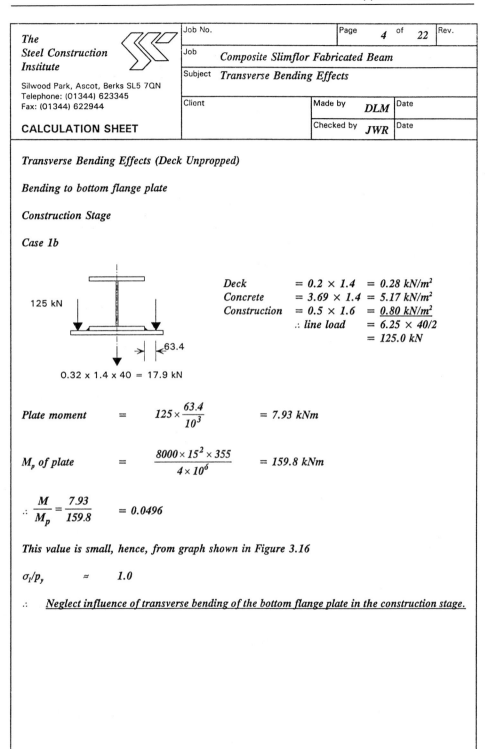

$$\begin{aligned}
Deck &= 0.2 \times 1.4 &&= 0.28 \ kN/m^2 \\
Concrete &= 3.69 \times 1.4 &&= 5.17 \ kN/m^2 \\
Construction &= 0.5 \times 1.6 &&= \underline{0.80 \ kN/m^2} \\
&\therefore line \ load &&= 6.25 \times 40/2 \\
&&&= 125.0 \ kN
\end{aligned}$$

125 kN

63.4

0.32 x 1.4 x 40 = 17.9 kN

$$Plate \ moment \quad = \quad 125 \times \frac{63.4}{10^3} \qquad = 7.93 \ kNm$$

$$M_p \ of \ plate \quad = \quad \frac{8000 \times 15^2 \times 355}{4 \times 10^6} \qquad = 159.8 \ kNm$$

$$\therefore \quad \frac{M}{M_p} = \frac{7.93}{159.8} \quad = 0.0496$$

This value is small, hence, from graph shown in Figure 3.16

$$\sigma_l/p_y \qquad \approx \qquad 1.0$$

$\therefore$ *Neglect influence of transverse bending of the bottom flange plate in the construction stage.*

The Steel Construction Institute	Job No.		Page 5 of 22	Rev.
Silwood Park, Ascot, Berks SL5 7QN Telephone: (01344) 623345 Fax: (01344) 622944	Job *Composite Slimflor Fabricated Beam*			
	Subject *Lateral Torsional Buckling*			
CALCULATION SHEET	Client		Made by *DLM*	Date
			Checked by *JWR*	Date

Lateral Torsional Buckling (LTB) (Deck Unpropped)
Construction stage
Case 1a

				kN
Deck	$0.2 \times 1.4 \times 40/2$	=		*5.6*
Concrete	$3.69 \times 1.4 \times 40/2$	=		*103.3*
Construction	$0.5 \times 1.6 \times 40/2$	=		*16.0*
				124.9 kN

Beam wt $= 0.32 \times 40 \times 1.4 = 17.9 \ kN$
Total load $= 124.9 + 17.9 + 5.6 = 148.4 \ kN$

Section Properties

<u>*254 × 254 × 107 UC Grd S355*</u>

Area	$= 136.6 \ cm^2$	
I_x	$= 17510 \ cm^4$	
I_y	$= 5901 \ cm^4$	
J	$= 172 \ cm^4$	
T	$= 20.5 > 16.0 \ mm$	
$\therefore p_y$	$= 345 \ N/mm^2$	

Plate Area $= 1.5 \times 46.0$
$= 69.0 \ cm^2$

Total Area, A_T $= 136.6 + 69.0$
$= 205.6 \ cm^2$

$$\bar{y} = \frac{(266.7 + 15) \, 69.0}{2(136.6 + 69.0)} = 47.3$$

$$I_{xx} = (17510 + 136.6 \times 4.73^2) + \left[\frac{266.7}{20} + \frac{15}{20} - 4.73\right]^2 69.0$$

$$= 17510 + 3056 + 6039$$

$$= 26605 \ cm^4$$

$$I_{yy} = 5901 + 1.5 \times 46^3/12 \qquad\qquad r_{yy} = \left[\frac{18068}{205.6}\right]^{1/2} = 9.37 \ cm$$

$$= 18068 \ cm^4$$

$$I_{tf} = TB^2/12 = 2.05 \times 25.83^3/12 \quad = 2944 \ cm^4 \ (top \ flange)$$

$$I_{if} = 2944 + 1.5 \times 46^3/12 \quad = 15111 \ cm^4 \ (bottom \ flange \ and \ plate)$$

The Steel Construction Institute		Job No.		Page 6 of 22	Rev.
		Job	Composite Slimflor Fabricated Beam		
Silwood Park, Ascot, Berks SL5 7QN		Subject	Lateral Torsional Buckling Continued.		
Telephone: (01344) 623345 Fax: (01344) 622944		Client		Made by DLM	Date
CALCULATION SHEET				Checked by JWR	Date

Buckling Parameter, u

$$u = \left[\frac{4 \times S_x^2 \times \gamma}{A^2 h_s^2} \right]^{1/4}$$

$R_s \;=\; A\,p_y \;=\; 136.6 \times 345/10 \qquad\qquad\qquad = 4712.7 \; kN$

$R_p \;=\; A_p\,p_y \;=\; 69.0 \times 355/10 \qquad\qquad = 2449.5 \; kN$

$R_f \;=\; 25.83 \times 2.05 \times 345/10 \qquad\qquad = 1826.8 \; kN$

$R_w \;=\; R_s - 2\,R_f = 4712.7 - 2 \times 1826.8 \qquad = 1059.1 \; kN$

$R_s > R_p > R_w \;\therefore\; pna \; in \; bottom \; UC \; flange$

position of plastic neutral axis from top of flange plate

$$y_p \;=\; \frac{4712.7 - 2449.5}{2 \times 258.3 \times 345 / 10^3} \qquad = 12.7 \; mm$$

$$M_c \;=\; \frac{4712.7 \times 266.7}{2 \times 10^3} + \frac{2449.5 \times 15}{2 \times 10^3} - \frac{(4712.7 - 2449.5)^2}{4 \times 258.3 \times 345}$$

$$= \; 628.4 + 18.4 - 14.4$$

$$= \; 632.4 \; kNm$$

$$\therefore S_x \;=\; \frac{632.4 \times 10^3}{345} \qquad = 1833 \; cm^3$$

$\gamma \;=\; 1 - I_y / I_x = 1 - 18068/26605 \qquad = 0.32$

$h_s \;\approx\; (D - T/2) = 266.7 - 20.5/2 \qquad = 256.5$

$$\therefore u \;=\; \left[\frac{4 \times 1833^2 \times 0.32}{205.6^2 \times 25.65^2} \right]^{1/4}$$

$$= \; 0.627$$

$$N \;=\; \frac{I_{cf}}{I_{cf} + I_{tf}} = \frac{2944}{2944 + 15111} = 0.163$$

$N < 0.5 \;\therefore\; \psi = 1.0 \; (2 \times 0.163 - 1) = -0.674$

		Job No.			Page	7	of	22	Rev.

The
Steel Construction
Institute

Silwood Park, Ascot, Berks SL5 7QN
Telephone: (01344) 623345
Fax: (01344) 622944

CALCULATION SHEET

Job   *Composite Slimflor Fabricated Beam*

Subject   *Lateral Torsional Buckling Continued.*

Client

Made by   *DLM*   Date

Checked by   *JWR*   Date

$$x \quad = 0.566 \, h_s \left(\frac{A}{J} \right)^{1/2}$$

$$J \quad = 172 + \tfrac{1}{3} \, (1.5^3 \times 46.0) \qquad = 223.7 \; cm^4$$

$$x \quad = 0.566 \times 25.65 \left[\frac{205.6}{223.7} \right]^{1/2} \qquad = 13.9$$

$$\lambda \quad = \frac{L_E}{r_y} = \frac{8000}{93.7} \quad = 85.4$$

$$\frac{\lambda}{x} \quad = \frac{85.4}{13.9} \qquad = 6.14$$

$$v \quad = \left[\left(4 \times 0.163 \, (1 - 0.163) + \frac{6.14^2}{20} + 0.674^2 \right)^{1/2} - 0.674 \right]^{-1/2}$$

$$= 0.988$$

$$\lambda_{LT} \quad = n \, u \, v \, \lambda$$

$$\therefore \lambda_{LT} \quad = 1.0 \times 0.627 \times 0.988 \times 85.4$$

$$= 52.9$$

$$\therefore p_y \quad = 276 \; N/mm^2 \; (using \; p_y = 345 \; N/mm^2)$$

$$M_b \quad = \frac{1833 \times 276}{10^3} \quad = 505.9 \; kNm$$

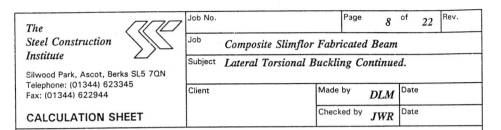

Job No.		Page	8	of	22	Rev.

The
*Steel Construction
Institute*

Silwood Park, Ascot, Berks SL5 7QN
Telephone: (01344) 623345
Fax: (01344) 622944

CALCULATION SHEET

Job   *Composite Slimflor Fabricated Beam*

Subject   *Lateral Torsional Buckling Continued.*

Client	Made by	*DLM*	Date
	Checked by	*JWR*	Date

Case 1a Continued (Torsion, using the simplified approach)

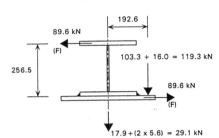

Total load $= 119.3 + 17.9 + 2 \times 5.6$

$\qquad = 148.4 \ kN$

$M_x = 148.4 \times 8/8 = 148.4 \ kNm$

$F = \dfrac{119.3 \times 192.6}{256.5} = 89.6 \ kN$

$M_y = 89.6 \times 8/8 = 89.6 \ kNm$

$M_{cy} = \dfrac{2.05 \times 25.83^2}{4} \times \dfrac{345}{10^3} = 118.0 \ kNm$

$\dfrac{M_x}{M_b} + \dfrac{M_y}{M_{cy}} \leq 1.0$

$\dfrac{148.4}{505.9} + \dfrac{89.6}{118.0} = 1.05 > 1.0 \qquad Say \ OK \qquad$ (*see note below*)

∴ *Beam satisfactory for out-of-balance loads combined with lateral torsional buckling.*

NB: The above approach to torsion is approximately 25% conservative compared to the rigorous approach given in Reference 9.

The *Steel Construction Institute*	Job No.		Page 9 of 22	Rev.
	Job Composite Slimflor Fabricated Beam			
Silwood Park, Ascot, Berks SL5 7QN	Subject Composite Stage (Transverse bending effects)			
Telephone: (01344) 623345 Fax: (01344) 622944	Client		Made by DLM	Date
CALCULATION SHEET			Checked by JWR	Date

Case 1b

Check for LTB

$$bending\ moment = (124.9 \times 2 + 17.9) \times \frac{8}{8} = 267.7\ kNm$$

from Sheet 7, M_b = 505.9 kNm

As <u>267.7 kNm < 505.9 kNm</u> the beam is satisfactory for LTB.

Composite Stage (Transverse bending effects)

Case 2a

301.6 kN

0.2 kN/m² Deck
3.69 kN/m² Concrete
0.25 kN/m² Services & Finishes
4.14 kN/m² × 1.4   =   5.8 Dead
5.80 kN/m² × 1.6   =   <u>9.28</u> Imposed
15.08 kN/m²

$$0.32 \times 1.4 \times 40 = 17.9\ kN$$

Eccentric load	=	$15.08 \times 40/2$	$= 301.6\ kN$
Total factored load	=	$2 \times 301.6 + 17.9$	$= 621.1\ kN$
Plate moment	=	$301.6 \times 63.4/10^3$	$= 19.12\ kNm/span$
M_p of plate	=	$\dfrac{8000 \times 15^2}{4 \times 10^6} \times 355$	$= 159.8\ kNm/span$

$$\therefore \frac{M}{M_p} = \frac{19.12}{159.8} = 0.12$$

from Figure 3.16 limit of $\sigma_l/p_y \approx 0.97$

∴ maximum value of $\sigma_l = 0.97 \times 355 = 344\ N/mm^2$

At the ultimate limit state for the composite section the value of p_y in the bottom flange plate will be limited to 344 N/mm²

∴ $R_p = 69.0 \times 344/10 = 2373.6\ kN$

		Job No.			Page 10	of 22	Rev.

The
Steel Construction
Institute

Silwood Park, Ascot, Berks SL5 7QN
Telephone: (01344) 623345
Fax: (01344) 622944

CALCULATION SHEET

Job	*Composite Slimflor Fabricated Beam*
Subject	*Deflection - Construction Stage Continued.*

Client		Made by	*DLM*	Date
		Checked by	*JWR*	Date

Construction Stage

Deflection ~ Case 1b

W $= (0.32 + 3.69 + 0.2) \, 40 = 168.4 \, kN$

Note: *Construction load not included for deflection purposes.*

$$\delta_c \quad = \frac{5 \times 168.4 \times 8000^3}{384 \times 205 \times 26605 \times 10^4} \quad = 20.6 \, mm \, (L/388) \quad OK$$

Construction stage deflection of 20.6 mm is considered as satisfactory.

		Job No.		Page _11_ of _22_	Rev.

The Steel Construction Institute

Silwood Park, Ascot, Berks SL5 7QN
Telephone: (01344) 623345
Fax: (01344) 622944

CALCULATION SHEET

Job	Composite Slimflor Fabricated Beam
Subject	Composite Stage

Client		Made by	_DLM_	Date
		Checked by	_JWR_	Date

Composite Stage

Full Shear Connection

since, $D_d \geq D$, *depth of concrete in compression* = D_s

$\therefore R_c = 0.45 f_{cu} B_e (D_s + D_d + T - D)$

$= 0.45 \times 30 \times 2000 (127 + 225 + 20.5 - 266.7)/10^3 = 2856.6 \ kN$

Shear Stud Connectors

19 dia $\times$ 70 mm high

$Q_k = 87 \ kN$ *(Shear connectors values are given in Reference 3)*

for LWC, $Q_k = 87 \times 0.9 = 78.3 \ kN$

Q_p *for positive moment regions* = $0.8 \times 78.3 = 62.6 \ kN$

Partial Shear Connection

Assume, minimum shear connection permitted in BS 5950: Part 3: Section 3.1.

$= 0.4 R_c$ *(note: 0.4 R_c, OK for spans up to 10 m)*

$= 0.4 \times 2856.6 = 1142.6 \ kN$

Number of shear connectors required $= \dfrac{1142.6}{62.6} = $ *18.3 say 19*
(per half span)

$\therefore$ *partial shear connection force,* R_q *becomes*

19 $\times$ 62.6 $= 1189.4 \ kN = R_q$

$R_q = 1189.4 \ kN, \ R_s = 4712.7 \ kN, \ R_p = 2373.6 \ kN, \ R_w = 1059.1 \ kN$

$(R_q + R_s) > R_p > (R_q + R_w)$

$\therefore$ *pna in bottom UC flange.*

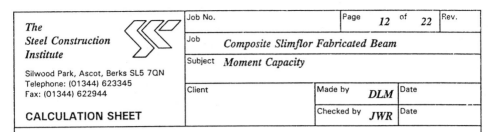

Job No.			Page 12 of 22	Rev.

The Steel Construction Institute

Silwood Park, Ascot, Berks SL5 7QN
Telephone: (01344) 623345
Fax: (01344) 622944

CALCULATION SHEET

Job	*Composite Slimflor Fabricated Beam*
Subject	*Moment Capacity*

Client		Made by	DLM	Date
		Checked by	JWR	Date

Moment Capacity

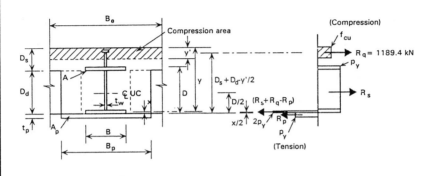

$$y' = \frac{1189.4 \times 10^3}{0.45 \times 30 \times 2000} = 44.1 \ mm$$

Position of pna

$$x = \frac{(4712.7 + 1189.4 - 2373.6) \, 10^3}{2 \times 258.3 \times 345} = 19.8 \ mm$$

pna in bottom UC flange.

$$M_c = \frac{1}{2} (R_s D + R_p t_p) + R_q (D_s + D_d - y'/2) - (R_s + R_q - R_p) \, x/2$$

$$= 1/2 (R_s D + R_p t_p) + R_q \left[D_s + D_d - \frac{R_q}{0.9 f_{cu} B_e} \right] - \frac{(R_s + R_q - R_p)^2}{4 B p_y}$$

$$= \frac{1}{2 \times 10^3} (4712.7 \times 266.7 + 2373.6 \times 15) + \frac{1189.4}{10^3} \left[127 + 225 - \frac{44.1}{2} \right]$$

$$- \frac{(4712.7 + 1189.4 - 2373.6)^2}{4 \times 258.3 \times 345}$$

$$= 646.2 + 392.4 - 34.9$$

$$= 1003.7 \ kNm$$

The *Steel Construction Institute* Silwood Park, Ascot, Berks SL5 7QN Telephone: (01344) 623345 Fax: (01344) 622944 **CALCULATION SHEET**	Job No.		Page *13*	of *22*	Rev.
	Job	*Composite Slimflor Fabricated Beam*			
	Subject	*Moment Capacity Continued.*			
	Client		Made by *DLM*	Date	
			Checked by *JWR*	Date	

Factored Bending Moment

$$w \quad = \quad (4.21 + 0.25) \, 1.4 + (5.8 \times 1.6) \qquad = 15.52 \; kN/m^2$$

$$W \quad = \quad 15.52 \times 8 \times 5 \quad = 620.8 \; kN$$

$$Moment \; = 620.8 \times 8/8 \quad = 620.8 \; kNm$$

$\therefore 620.8 \; kNm \; < \; 1003.7 \; kNm \qquad OK$

Moment capacity satisfactory using 40% shear connection.

Shear Connector Spacing

Beam span $= 8000 \; mm$

No. studs $= 38 \; per \; beam$

$$Stud \; spacing \quad = \quad \frac{8000}{38} \quad = 210 \; mm, \; Try \; 200 \; mm$$

$\therefore$ *Centre line of column to 1st stud position*

$$= \frac{8000 - 37 \times 200}{2}$$

$$= 300 \; mm$$

	Job No.			Page	14	of	22	Rev.

The
Steel Construction
Institute

Silwood Park, Ascot, Berks SL5 7QN
Telephone: (01344) 623345
Fax: (01344) 622944

CALCULATION SHEET

Job   *Composite Slimflor Fabricated Beam*

Subject   *Serviceability Stresses*

Client		Made by	DLM	Date
		Checked by	JWR	Date

Serviceability Stresses

Composite section ~ second moment of area
Modular ratio for LWC, $\alpha_e = 15$ (⅔ short term + ⅓ long term)

Elastic neutral axis, y_e

assume ena lies in zone 2, for $D > D_d$

The concrete below the decking is neglected and zone 2 has been assumed for the ena position. This means that the cracked and uncracked section properties are the same for this case.

$$C_1 \;=\; 127 + 225 - \frac{266.7}{2} \qquad = 218.7 \; mm$$

$$C_2 \;=\; 127 + 225 + \frac{15}{2} \qquad = 359.5 \; mm$$

$$y_e \;=\; \frac{\dfrac{2000}{2 \times 15}(127 + 225 - 266.7)^2 + 136.6 \times 10^2 \times 218.7 + 69.0 \times 10^2 \times 359.5}{\dfrac{2000}{15} + (127 + 225 - 266.7) + 205.6 \times 10^2}$$

$$= \; 186.4 \; mm$$

$$I_{xx} \;=\; \frac{B_e}{\alpha_e}(D_s + D_d - D)\left[\frac{(D_s + D_d - D)^2}{12} + k_1{}^2\right] + I_x\,(uc) + Ak_2{}^2 + A_p\,k_3{}^2$$

where:

$$k_1 \;=\; y_e - (D_s + D_d - D)/2) \;=\; 186.4 - (127 + 225 - 266.7)/2 \;=\; 143.8 \; mm$$

$$k_2 \;=\; D_s + D_d - D/2 - y_e \;=\; 127 + 225 - \frac{266.7}{2} - 186.4 \;=\; 32.3 \; mm$$

$$k_3 \;=\; D_s + D_d + t_p/2 - y_e \;=\; 127 + 225 + \frac{15}{2} - 186.4 \;=\; 173.1 \; mm$$

The		Job No.			Page	15	of	22	Rev.
Steel Construction Institute		Job	*Composite Slimflor Fabricated Beam*						
Silwood Park, Ascot, Berks SL5 7QN		Subject	*Serviceability Stresses Continued.*						
Telephone: (01344) 623345 Fax: (01344) 622944		Client			Made by	*DLM*		Date	
CALCULATION SHEET					Checked by	*JWR*		Date	

$$I_{xx} = \frac{2000}{15}(127 + 225 - 266.7)\left[\frac{(127 + 225 - 266.7)^2}{12} + 143.8^2\right]$$

$$+\ 17510 \times 10^4 + 136.6 \times 10^2 \times 32.3^2 + 69.0 \times 10^2 \times 173.1^2$$

$$=\ 638 \times 10^6\ mm^4$$

$$Z_x\ (concrete)\ =\ \frac{638 \times 10^6 \times 15}{18.64 \times 10^4}\quad =\quad 51341\ cm^3$$

$$Z_{xt}\ (steel)\quad =\ \frac{63800}{12.7 + 22.5 + 1.5 - 18.64}\quad =\quad 3533\ cm^3$$

$$Z_{xc}\ =\ \frac{63800}{10.11}\ =\ 6311\ cm^3$$

Steel Section only

$Z_x\ (Comp)\qquad = 26605/18.07\quad =\qquad 1472\ cm^3$

$Z_x\ (tension)\qquad = 26605/10.1\quad =\qquad 2634\ cm^3$

Loading (unfactored)

To steel beam only
Deck, Concrete and own wt of steel = 4.21 kN/m²

$W\ =\ 4.21 \times 8 \times 5 = 168.4\ kN$ $\qquad$ *Moment = 168.4 × 8/8 = 168.4 kNm*

To composite section
Services & Imposed $\qquad = \qquad 0.25 + 5.8 \quad = \qquad 6.05\ kN/m^2$

$W\ =\ 6.05 \times 8 \times 5 = 242.0\ kN$ $\qquad$ *Moment = 242.0 × 8/8 = 242.0 kNm*

Steel Stresses

Compression

Steel beam $\qquad 168.4 \times 10^3/1472\ = 114.4\ N/mm^2$

		Job No.		Page 16 of 22	Rev.

The
Steel Construction
Institute

Silwood Park, Ascot, Berks SL5 7QN
Telephone: (01344) 623345
Fax: (01344) 622944

CALCULATION SHEET

Job   *Composite Slimflor Fabricated Beam*

Subject   *Transverse Plate Bending - Elastic Condition*

Client

Made by   *DLM*   Date

Checked by   *JWR*   Date

Tension

Steel beam $168.4 \times 10^3/2634$ $= 63.9 \ N/mm^2$

Composite Section

Compression $242.0 \times 10^3/6311$ $= 38.3 \ N/mm^2$
Tension $242.0 \times 10^3/3533$ $= 68.5 \ N/mm^2$

Combined Steel Stresses

Compression $= 114.4 + 38.3 = 152.7 \ N/mm^2 < 345 \ N/mm^2$ *OK*
Tension $= 63.9 + 68.5 = 132.4 \ N/mm^2 < 355 \ N/mm^2$ *OK*

Concrete Stress

Compression $= 242.0 \times 10^3/51341$
    $= 4.7 \ N/mm^2 < f_{cu}/2 = 15 \ N/mm^2$ *OK*

		Job No.		Page	17	of	22	Rev.
The *Steel Construction* *Institute* Silwood Park, Ascot, Berks SL5 7QN Telephone: (01344) 623345 Fax: (01344) 622944		Job	*Composite Slimflor Fabricated Beam*					
		Subject	*Transverse Plate Bending - Elastic Conditions*					
		Client		Made by	*DLM*	Date		
CALCULATION SHEET				Checked by	*JWR*	Date		

Transverse Plate Bending - Elastic Condition

Elastic condition

Longitudinal stress, σ_t

σ_t *from Sheet 16 = 132.4 N/mm²*

plate bending

$$kN/m^2$$

Deck	= 0.20
Concrete	= 3.69

$$3.89 \times 40/2 = 77.8 \ kN$$

Imposed	= 5.8
Services	= 0.25

$$6.05 \times \frac{40}{2} = 121.0 \ kN$$

∴ *Flange plate line load*   =   *77.8 + 121.0 kN*

=   *198.8 kN*

plate bending   =   *198.8 × 63.4/10³ = 12.6 kNm*

plate modulus, Z_e

$$Z_e \ = \ \frac{1000 \times 15^2}{6 \times 10^3} \quad = 37.5 \ cm^3 \ per \ m$$

per span = 37.5 × 8   *= 300 cm³*

$$\sigma_n \ = \ \frac{12.6 \times 10^6}{300 \times 10^3} \quad = 42.0 \ N/mm^2$$

Combined stress must be less than or equal to p_y

i.e. $(\sigma_n^2 + \sigma_{t_o}^2 + \sigma_n \ \sigma_t)^{\frac{1}{2}} \not> p_y$

$(42.0^2 + 132.4^2 + 42.0 \times 132.4)^{\frac{1}{2}}$

=   *157.7 N/mm² < 355 N/mm² -*   *Serviceability stresses satisfactory.*

	Job No.			Page 18 of 22	Rev.
The **Steel Construction Institute**	Job		*Composite Slimflor Fabricated Beam*		
Silwood Park, Ascot, Berks SL5 7QN	Subject		*Deflection - Imposed*		
Telephone: (01344) 623345 Fax: (01344) 622944	Client		Made by **DLM**	Date	
CALCULATION SHEET			Checked by **JWR**	Date	

Deflection

Imposed

$$W = 5.8 \times 40 = 232.0 \ kN$$

$$\delta_i = \frac{5 \times 232.0 \times 8000^3}{384 \times 205 \times 638 \times 10^6} = 11.8 \ mm$$

Services & Finishes

$$\delta = 11.8 \times 0.25/5.8 = \underline{0.5 \ mm}$$
$$12.3 \ mm$$

As partial shear connection has been used for the stud connector design the deflection will have to be increased due to the slippage at the stud connector.

$$\textit{For unpropped construction} = \delta_c + 0.3\left[1 - \frac{N_a}{N_p}\right](\delta_s - \delta_c)$$

δ_s = *steel beam acting alone*

δ_c = *composite beam with full shear connection.*

$\dfrac{N_a}{N_p}$ = *0.4 (used for number of shear stud connectors see Sheet 11).*

$$\delta_s = \frac{5 \times 232.0 \times 8000^3}{384 \times 205 \times 26605 \times 10^4} = 28.4$$

$$\delta = 11.8 + 0.3 \ (1 - 0.4) \ (28.4 - 11.8)$$

$$= 11.8 + 3.0 = \underline{14.8 \ mm} \ (L/540 < L/360) \qquad\qquad OK$$

		Job No.		Page *19* of *22*	Rev.

The
Steel Construction
Institute

Silwood Park, Ascot, Berks SL5 7QN
Telephone: (01344) 623345
Fax: (01344) 622944

CALCULATION SHEET

Job	*Composite Slimflor Fabricated Beam*
Subject	*Deflection - Continued*

Client		Made by	*DLM*	Date
		Checked by	*JWR*	Date

Deflection Summary

Loading	Item	Deflection
0.2	*Deck*	
3.69	*Concrete*	
0.32	*Beam wt*	
4.21 kN/m²		*20.6 mm*
0.25 kN/m²	*Services & Finishes*	
5.80 kN/m²	*Imposed*	
6.05 kN/m²		*14.8 mm*

Total *35.4 mm*   *(L/226)*   *OK*

		Job No.			Page	**20**	of	**22**	Rev.

The
Steel Construction
Institute

Silwood Park, Ascot, Berks SL5 7QN
Telephone: (01344) 623345
Fax: (01344) 622944

CALCULATION SHEET

Job — *Composite Slimflor Fabricated Beam*

Subject — *Transverse Reinforcement*

Client — | Made by — **DLM** | Date
Checked by — **JWR** | Date

Transverse Reinforcement

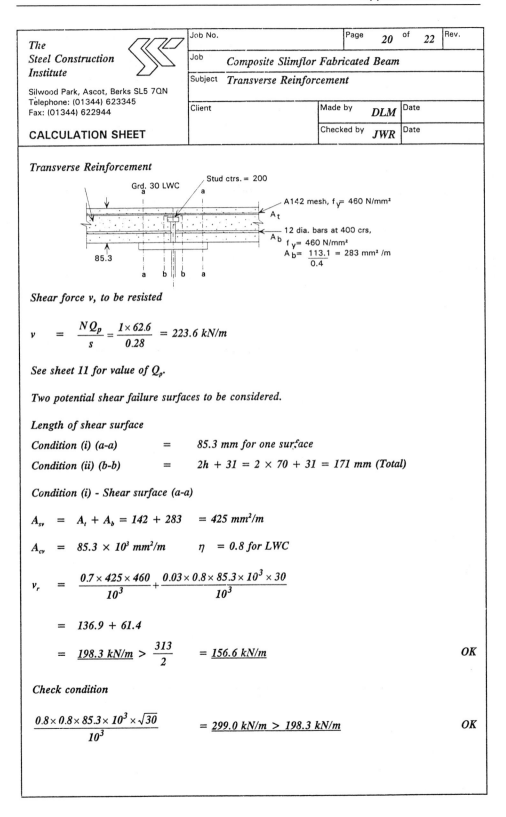

Shear force v, to be resisted

$$v \quad = \quad \frac{N Q_p}{s} = \frac{1 \times 62.6}{0.28} = 223.6 \ kN/m$$

See sheet 11 for value of Q_p.

Two potential shear failure surfaces to be considered.

Length of shear surface

Condition (i) (a-a)   =   *85.3 mm for one surface*

Condition (ii) (b-b)   =   *2h + 31 = 2 × 70 + 31 = 171 mm (Total)*

Condition (i) - Shear surface (a-a)

$$A_{sv} \ = \ A_t + A_b = 142 + 283 \ = 425 \ mm^2/m$$

$$A_{cv} \ = \ 85.3 \times 10^3 \ mm^2/m \qquad \eta \ = 0.8 \ for \ LWC$$

$$v_r \quad = \quad \frac{0.7 \times 425 \times 460}{10^3} + \frac{0.03 \times 0.8 \times 85.3 \times 10^3 \times 30}{10^3}$$

$$= \ 136.9 + 61.4$$

$$= \ \underline{198.3 \ kN/m} > \frac{313}{2} \ = \underline{156.6 \ kN/m} \qquad\qquad OK$$

Check condition

$$\frac{0.8 \times 0.8 \times 85.3 \times 10^3 \times \sqrt{30}}{10^3} \qquad = \underline{299.0 \ kN/m > 198.3 \ kN/m} \qquad\qquad OK$$

		Job No.			Page	21	of	22	Rev.
The *Steel Construction* *Institute*		Job	*Composite Slimflor Fabricated Beam*						
		Subject	*Transverse Reinforcement - Cont'd*						
Silwood Park, Ascot, Berks SL5 7QN Telephone: (01344) 623345 Fax: (01344) 622944		Client			Made by	*DLM*	Date		
CALCULATION SHEET					Checked by	*JWR*	Date		

Condition (ii)

Shear surface (b-b)

$$A_{sv} = 2\,A_b = 2 \times 283 = 566 \; mm^2/m$$

$$A_{cv} = 171 \times 10^3 \; mm^2/m$$

$$v_r = \frac{0.7 \times 566 \times 460}{10^3} + \frac{0.03 \times 0.8 \times 171 \times 10^3 \times 30}{10^3}$$

$$= 182.3 + 123.1$$

$$= 305.4 \; kN/m < 313 \; kN/m \qquad \textit{Except or alternatively increase the bar diameter.}$$

Check condition

$$0.8 \times 0.8 \times 171 \times 10^3 \times \sqrt{30} \,/\, 10^3 \; = 599.4 \, kN/m > \; 305.4 \, kN/m \qquad\qquad OK$$

∴   *Use A142 Mesh with 12 φ bars at 400 mm crs.*

	Job No.			Page 22 of 22	Rev.
The **Steel Construction Institute**	Job	*Composite Slimflor Fabricated Beam*			
Silwood Park, Ascot, Berks SL5 7QN Telephone: (01344) 623345 Fax: (01344) 622944	Subject	*Natural Frequency*			
CALCULATION SHEET	Client		Made by *DLM*	Date	
			Checked by *JWR*	Date	

Natural Frequency

Load = *Dead + 10% Imposed (not including partitions)*

Dead = *4.46 × 8 × 5 = 178.4 kN*

Imposed = *5.8 × 8 × 5 = 232 kN*

$\therefore W$ = *178.4 + 232/10 = 201.6 kN*

$$\delta_{nf} = \frac{5 \times 201.6 \times 8000^3}{384 \times 205 \times 638 \times 10^6}$$

= *10.3 mm*

$$Frequency = \frac{18}{(10.3)^{1/2}}$$

= *5.6 Hz > 4.0 Hz*   *OK*

Appendix C
Worked Example for Stub Girder Construction

A fully worked example of a stub girder based on a 15×12 m grid is presented. The design has been carried out in accordance with BS 5950: Part 3 and to a specified imposed load of 5 kN/m^2. In this example, the design is controlled by the combination of bending, shear and tension in the bottom chord adjacent to the most highly stressed opening, and by deflection criteria. The selected members (all grade S355 steel) were as follows:

- Secondary beams (12 m span) $457 \times 152 \times 60$ kg/m UB
- Bottom chord (stub girder) $305 \times 305 \times 198$ kg/m UC
- Top chord   T section $102 \times 203 \times 46$ kg/m cut from UC

Shear connectors are placed in pairs at 95 mm longitudinal spacing which determines the minimum length of stubs that are needed. Additional transverse reinforcement in the form of 16 mm diameter bars at 95 mm spacing is needed in the slab above the outer stubs.

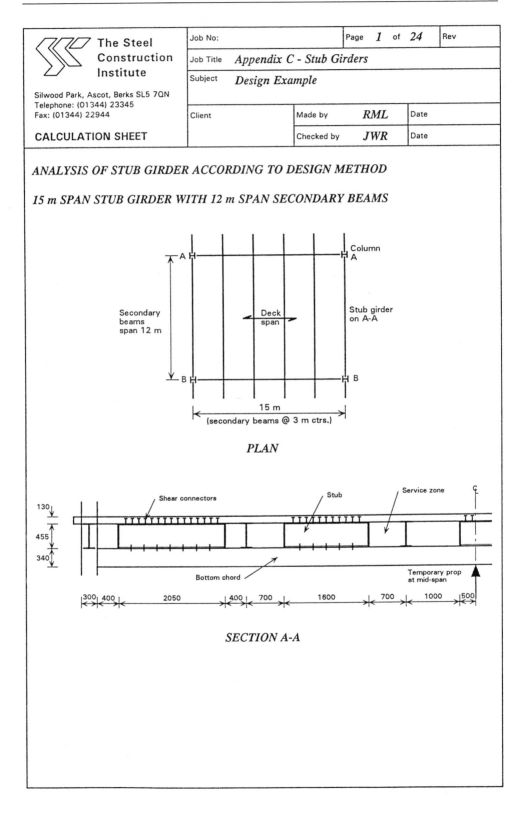

The Steel Construction Institute	Job No:		Page *1* of *24*	Rev
	Job Title	*Appendix C - Stub Girders*		
	Subject	*Design Example*		
Silwood Park, Ascot, Berks SL5 7QN Telephone: (01344) 23345 Fax: (01344) 22944	Client		Made by	*RML*	Date
CALCULATION SHEET			Checked by	*JWR*	Date

ANALYSIS OF STUB GIRDER ACCORDING TO DESIGN METHOD

15 m SPAN STUB GIRDER WITH 12 m SPAN SECONDARY BEAMS

Column A

Secondary beams span 12 m

Deck span

Stub girder on A-A

15 m
(secondary beams @ 3 m ctrs.)

PLAN

Shear connectors   Stub   Service zone

130
455
340

Bottom chord

Temporary prop at mid-span

300 400   2050   400 700   1600   700   1000 500

SECTION A-A

		Job No:		Page *2* of *24*	Rev

The Steel
Construction
Institute

Silwood Park, Ascot, Berks SL5 7QN
Telephone: (01344) 23345
Fax: (01344) 22944

CALCULATION SHEET

Job Title	*Appendix C - Stub Girders*
Subject	*Design Example*

Client	Made by	*RML*	Date
	Checked by	*JWR*	Date

DATA - GENERAL

Grid size	*12 m × 15 m*
Slab depth	*130 mm (90 min. fire resistance)*
Deck	*Re-entrant shape* *(50 mm deep × 1 mm thick)*
Concrete	*Lightweight, grade 30* *(f_{cu} = 30 N/mm²)*
Steel	*Grade S355 throughout* *(p_y = 345 N/mm² for t > 16 mm)*

Consider two construction cases:

a)   temporary props at 7.5 m from supports

b)   no temporary props, but additional T section as top chord

Bottom chord	*Choose 305 UC section*
Secondary beams	*Choose 457 UB section*
Columns	*Choose 305 UC section*
Maximum opening sizes:	*1000 mm × 455 mm (2N°)*
	700 mm × 455 mm (4N°)
	400 mm × 455 mm (4N°)

Total construction depth = 925 mm (span/16.2)

![logo] **The Steel Construction Institute**	Job No:	Page **3** of **24**	Rev
	Job Title *Appendix C - Stub Girders*		
	Subject *Design Example*		
Silwood Park, Ascot, Berks SL5 7QN Telephone: (01344) 23345 Fax: (01344) 22944	Client	Made by **RML**	Date
CALCULATION SHEET		Checked by **JWR**	Date

DATA - LOADING

Imposed loading	$4\ kN/m^2$
Partitions	$1\ kN/m^2$
Ceiling, Services, Raised Floor etc	$0.7\ kN/m^2$
Self Weight of Slab and Deck	$2.3\ kN/m^2$
Self Weight of Beams	$0.4\ kN/m^2$ *(assumed)*
Construction Load (Imposed)	$0.5\ kN/m^2$

Factored Load During Construction

$$w_c = 1.4 \times (2.3 + 0.4) + 1.6 \times 0.5$$

$$= 4.6\ kN/m^2$$

Factored Load In-Service

$$w_u = 1.4\ (2.3 + 0.4) + 1.4 \times 0.7 + 1.6\ (1.0 + 4.0)$$

$$= 12.8\ kN/m^2$$

Point Load due to Secondary Beams

$$P_u = w_u \times 12 \times 3$$

Mid Span Moment:

$$M_u = 2\ P_u \times 7.5 - P_u \times 4.5 - P_u \times 1.5$$

$$= 9\ P_u\ kNm \qquad \qquad \textit{(see diagram on page 11)}$$

		Job No:			Page *4* of *24*		Rev
The Steel Construction Institute		Job Title	*Appendix C - Stub Girders*				
		Subject	*Design Example*				
Silwood Park, Ascot, Berks SL5 7QN Telephone: (01344) 23345 Fax: (01344) 22944		Client		Made by	*RML*		Date
CALCULATION SHEET				Checked by	*JWR*		Date

DESIGN OF SECONDARY BEAMS

The secondary beams (except along the column line AB) may be designed as continuous beams. However, a member length of 24 m may be excessive. Therefore, design as simply-supported. Try span: depth = 20 (where depth = beam + slab).

The beams size may be justified by reference to the SCI publication 055 `Composite Slabs and Beams with Steel Decking'. For this span and load, choose:

457 × 152 × 60 kg/m UB grade S355 steel

A reduction in beam weight to 52 kg/m is justified if the beams not on the column line are designed as continuous.

Choose 457 × 152 × 60 kg/m   UB as stubs
(ie. same section as for secondary beams)

![SCI logo] The Steel Construction Institute	Job No:		Page 5 of 24	Rev
	Job Title *Appendix C - Stub Girders*			
Silwood Park, Ascot, Berks SL5 7QN Telephone: (01344) 23345 Fax: (01344) 22944	Subject *Design Example*			
	Client	Made by *RML*	Date	
CALCULATION SHEET		Checked by *JWR*	Date	

CONSTRUCTION CONDITION

a)   *temporary prop at mid-span (i.e. 7.5 m from support)*

   Design bottom chord conservatively as propped cantilever

   Moment $\approx$ $w_c\, L^2/8$

            $=$ $4.6 \times 12 \times 7.5^2/8$

            $=$ *388.1 kNm*

Choose 305 $\times$ 305 UC $\times$ 198 kg/m grade S355 which is a `plastic' section. Moment capacity (using a design strength of steel of 345 N/mm² (t > 16 mm)) is:

      M_s $=$ $3440 \times 10^3 \times 345 \times 10^{-6}$

            $=$ *1186.8 kNm* $>$ *388.1 kNm*   *OK*

b)   *no temporary props, but use T section as top chord*

      Moment $=$ $9\, P_u$ $=$ $9 \times 4.6 \times 3 \times 12$ *(see page 3)*

            $=$ *1490 kNm*

But moment capacity of bottom chord = 1186.8 kNm

Consider T section of area $A_t = 55$ cm² $\approx$ area of web of bottom chord ($A_w = 54$ cm²). Plastic neutral axis lies in top flange.

Moment capacity $\approx A_t\, D_t\, p_y + A\, D/2\, p_y$

where D_t is the height of the T section above the top flange

	Job No:		Page *6* of *24*	Rev

The Steel Construction Institute

Silwood Park, Ascot, Berks SL5 7QN
Telephone: (01344) 23345
Fax: (01344) 22944

CALCULATION SHEET

Job Title   *Appendix C - Stub Girders*

Subject   *Design Example*

Client	Made by	*RML*	Date
	Checked by	*JWR*	Date

Choose T cut from 203 × 203 × 86 kg/m UC grade S355 so that upstand is less than height of slab minus top cover (in this case the T section is 111 mm high and 43 kg/m weight).

D_t ≈ 455 + 10 *(allowance for height of centroid of T section)* = 465 mm

$$M_c = 55 \times 10^2 \times 465 \times 345 \times 10^{-6}$$

$$+ \quad 252 \times 10^2 \times 340 \times 345 \times 10^{-6}$$

$$= \quad 882 + 2955 = 3837 \, kNm > 1490 \, kNm$$

This moment capacity exceeds the applied moment by a significant margin. As the bottom chord is assumed to resist the total shear and `Vierendeel' bending effects under factored in-service conditions, no further checks need be done at the construction stage.

	Job No:		Page **7** of **24**	Rev
The Steel Construction Institute	Job Title *Appendix C - Stub Girders*			
	Subject *Design Example*			
Silwood Park, Ascot, Berks SL5 7QN Telephone: (01344) 23345 Fax: (01344) 22944	Client	Made by	*RML*	Date
CALCULATION SHEET		Checked by	*JWR*	Date

ULTIMATE CONDITION

Point load P_u = $36 \times 12.8 = 461 \; kN$

Mid-span moment

$$M_u \quad = \quad 9 P_u \quad = \quad 4147 \; kNm \qquad (see \; page \; 3)$$

Moment capacity of bottom chord $M_s = 1186.8 \; kNm$

Compressive resistance of concrete slab (ignoring contribution of T section, conservatively).

Effective breadth of slab

$$B_e \quad = \quad L/4 \; \ngtr \; 0.8b$$

$$= \quad 15000/4 = 3750 \; \ngtr \; 12000 \times 0.8$$

$$R_c \quad = \quad 0.45 \, f_{cu} \, B_e \, D_{av}$$

D_{av} =   *equivalent depth of solid slab for re-entrant deck with ribs parallel to beam* $\approx 120 \; mm$

$$R_c \quad = \quad 0.45 \times 30 \times 3750 \times 120 \times 10^{-3}$$

$$= \quad 6075 \; kN$$

Tensile resistance of bottom chord

$$R_s \quad = \quad A \, p_y$$

$$= \quad 252 \times 10^2 \times 345 \times 10^{-3}$$

$$= \quad 8694 \; kN$$

The Steel Construction Institute	Job No:			Page *8* of *24*			Rev
	Job Title	*Appendix C - Stub Girders*					
	Subject	*Design Example*					
Silwood Park, Ascot, Berks SL5 7QN Telephone: (01344) 23345 Fax: (01344) 22944	Client			Made by	*RML*		Date
CALCULATION SHEET				Checked by	*JWR*		Date

Check moment capacity of stub girder assuming full shear connection.

Tensile resistance of web of bottom chord

$$R_w = 277 \times 19.2 \times 345 \times 10^{-3}$$
$$= 1835 \ kN$$

As $R_s > R_c > R_w$ plastic neutral axis lies in top flange of bottom chord

For simplicity, the plastic neutral axis is taken to be at the top of the top flange of the bottom chord.

Moment capacity, (about flange)

$$M_c = R_c \ D_c + R_s \ D/2$$

where D_c = height of centre of concrete slab above the top flange of bottom chord $\approx 455 + 130/2$ mm

$$M_c = 6075 \times (455 + 130/2) \times 10^{-3}$$

$$+ \quad 8694 \times 340/2 \times 10^{-3}$$

$$= \quad 3159.0 + 1478.0$$

$$= \quad 4637.0 \ kNm > 4147 \ kNm$$

This is acceptable. An additional conservative factor is that the span of the stub girder has been taken as from the centre of the columns (rather than the face).

Shear connection

Force to be transferred for full shear connection $= R_c = 6075 \ kN < R_s$

Design capacity of 19 mm dia shear connectors $\times$ 95 mm high in LWC (grade 30)

$$Q_p = 0.8 \times 0.9 \times 100$$
$$= 72 \ kN$$

No reduction for deck shape in this case

	Job No:		Page *9* of *24*	Rev
The Steel Construction Institute	Job Title	*Appendix C - Stub Girders*		
	Subject	*Design Example*		
Silwood Park, Ascot, Berks SL5 7QN Telephone: (01344) 23345 Fax: (01344) 22944	Client		Made by *RML*	Date
CALCULATION SHEET			Checked by *JWR*	Date

Number of shear connectors required in ½ span

$$N_{sc} \quad = \quad 6075/72 = 84.4, \; say \; 85$$

*Minimum spacing of shear connectors = 5φ (= 95 mm)
longitudinally and 4φ laterally in pairs*

Therefore, overall length of stubs needed

$$= \quad 84 \times 5 \times 19/2 = 3990 \; mm$$

Length available   =   2050 + 1600 = 3650 mm < 3990 mm

(ignoring middle stub)

Consider partial shear connection:

Minimum degree of shear connection

$$= \quad (L - 6)/10 \quad = (15 - 6)/10 = 0.9$$

Reduce number of shear connectors to 84.4 × 0.9 = 76

Overall length of stubs needed

$$= \quad 75 \times 5 \times 19/2 \quad = 3563 \; mm < 3650 \; mm \qquad\qquad OK$$

Check moment capacity of stub girder for partial shear connection

Longitudinal force transferred

$$R_q \quad = \quad 76 \times 72 \quad = 5472 \; kN$$

Resistance of web R_w = 1835 kN < R_q

The Steel Construction Institute	Job No:		Page *10* of *24*		Rev
	Job Title	*Appendix C - Stub Girders*			
Silwood Park, Ascot, Berks SL5 7QN Telephone: (01344) 23345 Fax: (01344) 22944	Subject	*Design Example*			
	Client		Made by	*RML*	Date
CALCULATION SHEET			Checked by	*JWR*	Date

Plastic neutral axis lies in top flange of bottom chord

$$M_c = R_q D_c + R_s D/2$$

$$= 5472 (455 + 130/2) \times 10^{-3} + 8694 \times 340/2 \times 10^{-3}$$

$$= 2845.4 + 1478.0$$

$$= 4323 \text{ kNm} > 4147 \text{ kNm} \qquad\qquad OK$$

Therefore partial shear connection is acceptable

Check shear capacity of bottom chord

Maximum shear force at support

$$V = 2 P_u = 2 \times 461 = 922 \text{ kN}$$

Shear capacity of web

$$V_u = 0.6 p_y A_v$$

$$= 0.6 \times 345 \times 19.2 \times 340 \times 10^{-3}$$

$$= 1351 \text{ kN}$$

$V < V_u$ *at the supports, which is acceptable.*

However as $V > 0.6 V_u$, *it is necessary to consider the influence of combined moment and shear adjacent to the first stub (see page 13)*

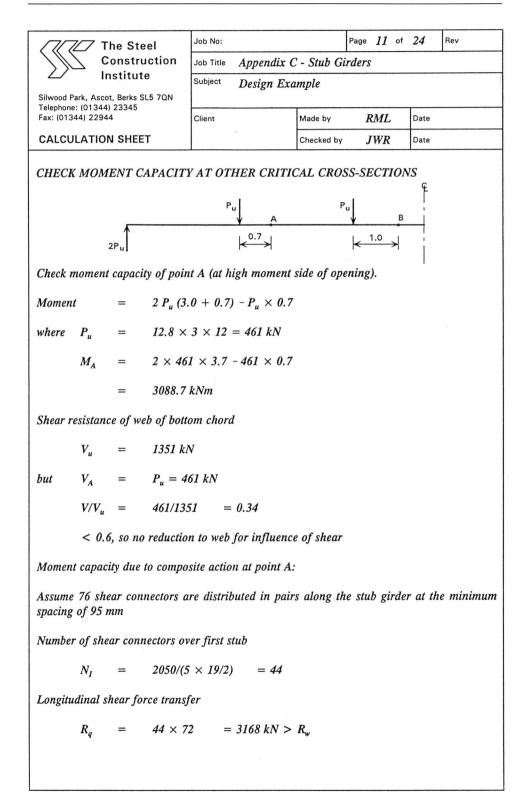

CHECK MOMENT CAPACITY AT OTHER CRITICAL CROSS-SECTIONS

Check moment capacity of point A (at high moment side of opening).

$$Moment \quad = \quad 2\,P_u\,(3.0 + 0.7) - P_u \times 0.7$$

$$where \quad P_u \quad = \quad 12.8 \times 3 \times 12 = 461\ kN$$

$$M_A \quad = \quad 2 \times 461 \times 3.7 - 461 \times 0.7$$

$$= \quad 3088.7\ kNm$$

Shear resistance of web of bottom chord

$$V_u \quad = \quad 1351\ kN$$

$$but \quad V_A \quad = \quad P_u = 461\ kN$$

$$V/V_u \quad = \quad 461/1351 \quad = 0.34$$

$$< 0.6,\ so\ no\ reduction\ to\ web\ for\ influence\ of\ shear$$

Moment capacity due to composite action at point A:

Assume 76 shear connectors are distributed in pairs along the stub girder at the minimum spacing of 95 mm

Number of shear connectors over first stub

$$N_1 \quad = \quad 2050/(5 \times 19/2) \quad = 44$$

Longitudinal shear force transfer

$$R_q \quad = \quad 44 \times 72 \quad = 3168\ kN > R_w$$

The Steel Construction Institute	Job No:		Page *12* of *24*	Rev
	Job Title	*Appendix C - Stub Girders*		
Silwood Park, Ascot, Berks SL5 7QN Telephone: (01344) 23345 Fax: (01344) 22944	Subject	*Design Example*		
	Client		Made by *RML*	Date
CALCULATION SHEET			Checked by *JWR*	Date

Moment capacity at point A

$$M_c = R_q D_c + R_s D/2$$

$$= 3168 \, (455 + 130/2) \times 10^{-3} + 1478.0$$

$$= 3125.4 \, kNm$$

This just exceeds the applied moment of 3088.7 kNm.

Moment capacity at point B is as calculated previously   = 4323.4 kNm

This exceeds the applied moment

Number of shear connectors over second stub

$$N_2 = 76 - 44 \qquad = 32$$

Nominal spacing of shear connectors over central stub, singly at 5ϕ spacing

$$N_3 = 1000/(5 \times 19) \qquad = 10$$

Total number of shear connectors in span

$$= 76 \times 2 + 10 \qquad = 162$$

		Job No:			Page *13* of *24*	Rev
The Steel Construction Institute		Job Title	*Appendix C - Stub Girders*			
		Subject	*Design Example*			
Silwood Park, Ascot, Berks SL5 7QN Telephone: (01344) 23345 Fax: (01344) 22944		Client		Made by	*RML*	Date
CALCULATION SHEET				Checked by	*JWR*	Date

VIERENDEEL MOMENT CAPACITY OF BOTTOM CHORD

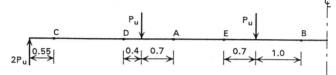

Adjacent to support at point C:

Vertical shear force adjacent to first stub

$$V \quad = \quad 2\,P_u = 2 \times 461 \quad = 922\ kN$$

Vierendeel moment transfer across opening adjacent to first stub, assuming point of contraflexure at location of secondary beam

$$M_v \quad = \quad 922 \times 0.4 \quad = 368.8\ kNm$$

Moment capacity of steel section $\quad M_s \quad = 1186.8\ kNm$

but $\quad V/V_u \quad = \quad 922/1351 \quad = 0.68 > 0.6$

By inspection, even ignoring web effectiveness in bending

$$M_s \quad >> \quad M_v\ at\ point\ C$$

Check point D:

Tensile force in bottom chord at point D

$$= \quad force\ transferred\ by\ shear\ connectors\ over\ first\ stub$$

$$T \quad = \quad R_q \quad = 3168\ kN \quad\quad (see\ page\ 11)$$

$$V/V_u \quad = \quad 0.68,\ as\ above$$

The Steel Construction Institute	Job No:		Page *14* of *24*	Rev
	Job Title	*Appendix C - Stub Girders*		
	Subject	*Design Example*		

Silwood Park, Ascot, Berks SL5 7QN
Telephone: (01344) 23345
Fax: (01344) 22944

Client	Made by *RML*	Date
CALCULATION SHEET	Checked by *JWR*	Date

Effective thickness of web reduced due to influence of shear

$$t_e \quad = \quad t_w \sqrt{1-(V/V_u)^2}$$

$$= \quad 19.2 \sqrt{1-0.68^2} = 14.0\,m$$

Effective properties of section

$$A_{eff} \quad = \quad 25200 - (19.2 - 14.0)\,(340 - 2 \times 31.4) = 23,759\,mm^2$$

$$M_{s,eff} \quad = \quad M_s - (19.2 - 14.0) \times (340 - 2 \times 31.4)^2 \times 10^{-6}/4$$

$$= \quad 1152.3\,kNm$$

Reduced moment capacity of steel section under the influence of shear and tension

$$\frac{M_{s,red}}{M_{s,eff}} \quad = \quad 1 - \frac{T}{A_{eff}\,p_y}$$

$$= \quad 1 - \frac{3168 \times 10^3}{23759 \times 345}$$

$$= \quad 1 - 0.39 \qquad = 0.61$$

$$M_{s,red} \quad = \quad 0.61 \times 1152.3 \qquad = 703\,kNm > M_v$$

Check point A:

$$V \quad = \quad P_u \quad = 461\,kN$$

$$M_v \quad = \quad 461 \times 0.7 \quad = 322.7\,kNm$$

As the section is not reduced by shear, it is adequate, by inspection

	Job No:			Page *15* of *24*	Rev
The Steel Construction Institute	Job Title	*Appendix C - Stub Girders*			
	Subject	*Design Example*			
Silwood Park, Ascot, Berks SL5 7QN Telephone: (01344) 23345 Fax: (01344) 22944	Client		Made by	*RML*	Date
CALCULATION SHEET			Checked by	*JWR*	Date

Check point E:

Tensile force in bottom chord at point E

$$= \text{Total force transferred by shear connectors across first and second stubs}$$

$$T \quad = \quad 76 \times 72 \quad = 5472 \ kN$$

No reduction for shear in this case

$$\frac{M_{s,red}}{M_s} \quad = \quad 1 - \frac{5472}{8694} = 0.37$$

$$M_{s,red} \quad = \quad 0.37 \times 1186.8 \quad = 439.1 \ kNm$$

$$M_v \quad = \quad 461 \times 0.7 \quad = 322.7 \ kNm < 439.1 \ kNm \qquad OK$$

Check point B:

For openings in `zero shear' zone, assume a shear force of ¼ × maximum shear representing the effect of unequal loading on adjacent spans.

$$V \quad = \quad 2 \times 461/4 \quad = 230.5 \ kN$$

Vierendeel moment transfer

$$M_v \quad = \quad 230.5 \times 1.0 \ = 230.5 \ kNm$$

But $M_{s,red} \ = \ 439.1 \ kNm \ (as \ for \ point \ E) > M_v$

Therefore the width of all the openings is acceptable. The width of the openings could be increased, but it is not possible to locate all the required shear connectors over the stub.

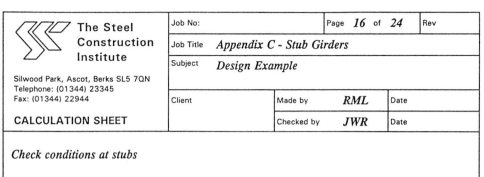

The Steel Construction Institute	Job No:	Page *16* of *24*	Rev
	Job Title *Appendix C - Stub Girders*		
Silwood Park, Ascot, Berks SL5 7QN Telephone: (01344) 23345 Fax: (01344) 22944	Subject *Design Example*		
	Client	Made by *RML*	Date
CALCULATION SHEET		Checked by *JWR*	Date

Check conditions at stubs

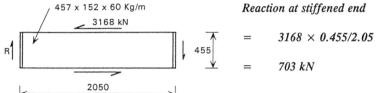

457 x 152 x 60 Kg/m

3168 kN

R

455

2050

Reaction at stiffened end

$$= \quad 3168 \times 0.455/2.05$$

$$= \quad 703 \ kN$$

Shear stress in web of stub

$$= \quad 3168 \times 10^3/(2050 \times 8.0)$$

$$= \quad 193 \ N/mm^2$$

Limiting shear stress $= 0.6 \, p_y$

$$= \quad 0.6 \times 355 \quad = 213 \ N/mm^2 \qquad\qquad just \ OK$$

Design web stiffeners for reaction of 703 kN

Use 20 mm thick × 150 mm wide × 455 mm deep plates welded to the ends of the stubs by 8 mm fillet welds

Compressive stress $= \quad 703 \times 10^3/(20 \times 150) \quad = 234 \ N/mm^2$

The stiffener should be checked as in BS 5950 Part 1, but the above size is acceptable

At second stub, shear stress in web

$$= \quad 2304 \times 10^3/(8.0 \times 1600)$$

$$= \quad 180 \ N/mm^2 \quad < 213 \ N/mm^2 \qquad\qquad OK$$

The Steel
Construction
Institute

Silwood Park, Ascot, Berks SL5 7QN
Telephone: (01344) 23345
Fax: (01344) 22944

Job No:		Page *17* of *24*	Rev
Job Title	*Appendix C - Stub Girders*		
Subject	*Design Example*		
Client		Made by	*RML*	Date
		Checked by	*JWR*	Date

CALCULATION SHEET

Vertical reaction with no stiffeners

$$= \quad \frac{2304 \times 10^3 \times 455}{(1600^2 \, / \, 6)}$$

$$= \quad 2457 \; N/mm$$

Vertical stress $=$ $\quad 2457/8.0 \quad = 307 \; N/mm^2$

Combined stress $\quad = \quad (307^2 + 3 \times 180^2)^{\frac{1}{2}}$
$\qquad\qquad\qquad = \quad 437 \; N/mm^2 > 355 \; N/mm^2$

Therefore use vertical stiffeners as previously. However, it may have been possible to justify the use of a heavier stub (e.g. 457 × 152 × 82 kg/m) although this section is actually 10 mm deeper than the chosen section.

Force transfer between the stubs and the bottom chord

Shear force at outer stub = 3168 kN
Use 10 mm fillet welds along flange tips and at end of stiffeners

Assume longitudinal shear is resisted by these fillet welds and vertical reactions are resisted by vertical stiffener welds

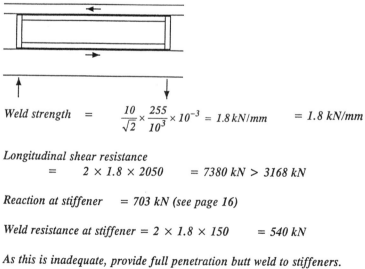

Weld strength $\quad = \quad \dfrac{10}{\sqrt{2}} \times \dfrac{255}{10^3} \times 10^{-3} = 1.8 \, kN/mm \qquad = 1.8 \; kN/mm$

Longitudinal shear resistance
$\qquad = \quad 2 \times 1.8 \times 2050 \qquad = 7380 \; kN > 3168 \; kN$

Reaction at stiffener $\quad = 703 \; kN \; (see \; page \; 16)$

Weld resistance at stiffener = 2 × 1.8 × 150 $\qquad = 540 \; kN$

As this is inadequate, provide full penetration butt weld to stiffeners.

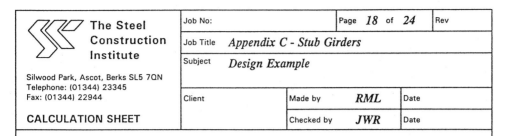

The Steel Construction Institute	Job No:		Page *18* of *24*	Rev
	Job Title	*Appendix C - Stub Girders*		
Silwood Park, Ascot, Berks SL5 7QN Telephone: (01344) 23345 Fax: (01344) 22944	Subject	*Design Example*		
CALCULATION SHEET	Client	Made by	*RML*	Date
		Checked by	*JWR*	Date

Crippling of web of bottom chord under stub stiffener

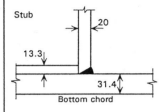

Bearing length on web of bottom chord (conservatively)

$$\ell_b \quad = \quad 20 + 13.3 \times 2.5 + 31.4 \times 2 \times 2.5 \quad = 210 \, mm$$

$$Bearing \; stress \quad = \quad \frac{704 \times 10^3}{210 \times 19.2} \quad = 175 \, N/mm^2 < p_y$$

Slenderness for buckling check

$$\lambda \quad = \quad 2.5 \, d/t = 2.5 \times 246/19.2 \; = 32$$

By inspection, web is OK

The Steel Construction Institute	Job No:	Page *19* of *24*	Rev
	Job Title *Appendix C - Stub Girders*		
	Subject *Design Example*		
Silwood Park, Ascot, Berks SL5 7QN Telephone: (01344) 23345 Fax: (01344) 22944	Client	Made by *RML*	Date
CALCULATION SHEET		Checked by *JWR*	Date

DEFLECTIONS

No checks are made on serviceability stresses in stub girders because a small degree of plasticity is not considered to affect deflections significantly.

Calculate imposed load deflections

Second moment ratio of steel to concrete, $\alpha_e = 15$

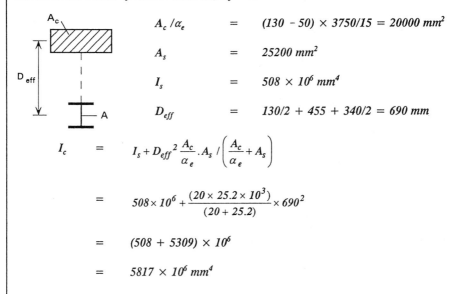

$$A_c / \alpha_e \quad = \quad (130 - 50) \times 3750/15 = 20000 \ mm^2$$

$$A_s \quad = \quad 25200 \ mm^2$$

$$I_s \quad = \quad 508 \times 10^6 \ mm^4$$

$$D_{eff} \quad = \quad 130/2 + 455 + 340/2 = 690 \ mm$$

$$I_c \quad = \quad I_s + D_{eff}^{\ 2} \frac{A_c}{\alpha_e} \cdot A_s \left/ \left(\frac{A_c}{\alpha_e} + A_s \right) \right.$$

$$= \quad 508 \times 10^6 + \frac{(20 \times 25.2 \times 10^3)}{(20 + 25.2)} \times 690^2$$

$$= \quad (508 + 5309) \times 10^6$$

$$= \quad 5817 \times 10^6 \ mm^4$$

Serviceability loading on beam

$$= \quad (4 + 1) \times 12 \qquad = 60 \ kN/m$$

Deflection due to pure bending

$$\delta_i \quad \approx \quad \frac{5}{384} \times \frac{wL^4}{EI_c}$$

$$= \quad \frac{5}{384} \times \frac{60 \times 15^4 \times 10^9}{205 \times 5817 \times 10^6}$$

$$= \quad 33.1 \ mm$$

The Steel Construction Institute	Job No:		Page *20* of *24*	Rev
	Job Title	*Appendix C - Stub Girders*		
Silwood Park, Ascot, Berks SL5 7QN Telephone: (01344) 23345 Fax: (01344) 22944	Subject	*Design Example*		
	Client		Made by *RML*	Date
CALCULATION SHEET			Checked by *JWR*	Date

Deflection due to Vierendeel bending

$$\delta_v \quad = \quad \frac{2}{3} \frac{V L_o^3}{E I_s} \qquad per\ opening$$

$$V \quad = \quad shear\ force\ at\ opening$$

$$L_o \quad = \quad length\ of\ opening\ (to\ point\ of\ contraflexure)$$

At first opening $\quad V = 2 P \quad = \quad 2 \times 12 \times 3 \times (4 + 1) = 360\ kN$

$$L_o \quad = \quad 400\ mm$$

$$\delta_v \quad = \quad \frac{2}{3} \times \frac{360 \times 400^3}{200 \times 508 \times 10^6} \quad = 0.15\ mm$$

For second opening $\quad V = P \quad = \quad 180\ kN$

$$L_o \quad = \quad 700\ mm$$

$$\delta_v \quad = \quad \frac{2}{3} \times \frac{180 \times 700^3}{200 \times 508 \times 10^6} \quad = 0.41\ mm$$

Total deflection due to Vierendeel bending over all openings $\approx 1.1\ mm$

Total deflection due to imposed loading

$$= \quad 33.1 + 1.1 \quad = 34.2\ mm$$

$$= \quad span\ /\ 439 \qquad OK$$

<table>
<tr><td colspan="2" rowspan="4">
⟨⟨⟨⟨ **The Steel
Construction
Institute**

Silwood Park, Ascot, Berks SL5 7QN
Telephone: (01344) 23345
Fax: (01344) 22944

CALCULATION SHEET
</td><td colspan="3">Job No:</td><td>Page *21* of *24*</td><td>Rev</td></tr>
<tr><td colspan="3">Job Title   *Appendix C - Stub Girders*</td><td colspan="2"></td></tr>
<tr><td colspan="3">Subject   *Design Example*</td><td colspan="2"></td></tr>
<tr><td>Client</td><td>Made by</td><td>*RML*</td><td colspan="2">Date</td></tr>
<tr><td colspan="2"></td><td></td><td>Checked by</td><td>*JWR*</td><td colspan="2">Date</td></tr>
</table>

TRANSVERSE REINFORCEMENT

Maximum longitudinal shear transfer at outer stubs

$$= \quad 2 \times 72 \times 10^3 /(5 \times 19) \quad = 1516\ kN/m$$

Shear stress on two shear planes through minimum slab depth

$$V \quad = \quad 1516/(2\ (130 - 50)) = 9.47\ N/mm^2$$

Shear resistance per unit length

$$V_c \quad = \quad 0.03 \times 0.8\,f_{cu}\,(D_s - D_p) + 0.7\,A_{sv}\,f_y + v_p$$

Use T16 bars at 95 mm spacing over stub and A142 mesh.
Place bars below head of studs

Note: Shear transfer is assumed to occur directly over the stub, which is conservative

$$A_s \quad = \quad \frac{\pi}{4} \times 16^2 \times \frac{1000}{95} + 142 \ = 2258\,mm^2$$

$$V_c \quad = \quad 0.03 \times 0.8 \times 30\ (130-50) + 0.7 \times 2258 \times 460 \times 10^{-3} + v_p$$

$$= \quad 57.6 + 727.1 + v_p$$

$$= \quad 784.7\ N/mm\ (or\ kN/m)$$

For 2 shear planes $V_c \quad = \quad 1569.4\ N/mm > 1516\ N/mm$

This is satisfactory, even ignoring the contribution of the decking, v_p. Curtail half of the bars at 1.0 m from the beam, and the remainder at 2 m.

The Steel Construction Institute	Job No:		Page **22** of **24**	Rev
	Job Title	*Appendix C - Stub Girders*		
Silwood Park, Ascot, Berks SL5 7QN Telephone: (01344) 23345 Fax: (01344) 22944	Subject	*Design Example*		
	Client		Made by	*RML*	Date
CALCULATION SHEET			Checked by	*JWR*	Date

DEFLECTION DUE TO SELF WEIGHT

a)   Propped beam

On removal of props reapply self weight to composite section. Take $\alpha_e = 25$ for long term loads

$$A_c/\alpha_e = \quad (130 - 50) \times 3750/25 \quad = 12000 \; mm^2$$

$$I_c = \quad 508 \times 10^6 + \frac{(12 \times 25.2 \times 10^3)}{(12 + 25.2)} \times 690^2$$

$$= \quad (508 + 3870) \times 10^6$$

$$= \quad 4378 \times 10^6 \; mm^4$$

Deflection on removal of props

Self weight loading   =   2.7 × 12   = 32.4 kN/m

$$\delta_{sw} \quad = \quad \frac{5}{384} \times \frac{32.4 \times 15^4 \times 10^9}{205 \times 4378 \times 10^6}$$

$$= \quad 23.8 \; mm$$

Vierendeel deflections are small (allow 0.7 mm)

Additional dead load deflections applied to composite section are due to a load of 0.7 kN/m² (ceiling + services)

Total deflection   =   24.5 + 5.7/5.0 × 34.2

$$= \quad 63.5 \; mm \; (span/236) \; This \; is \; probably \; just \; acceptable$$

The Steel Construction Institute	Job No:		Page *23* of *24*		Rev
	Job Title	*Appendix C - Stub Girders*			
	Subject	*Design Example*			
Silwood Park, Ascot, Berks SL5 7QN Telephone: (01344) 23345 Fax: (01344) 22944	Client		Made by	*RML*	Date
CALCULATION SHEET			Checked by	*JWR*	Date

b)   *Unpropped beam with top chord*

Area of top chord   =   *5500 mm^2*

Modified second moment of area of bottom and top chords is:

$$I_s \quad = \quad 508 \times 10^6 + \frac{(5.5 \times 25.2 \times 10^3)}{(5.5 + 25.2)} \times 690^2$$

$$= \quad (508 + 2149) \times 10^6$$

$$= \quad 2657 \times 10^6 \; mm^2$$

$$\delta_{sw} \quad = \quad 23.8 \times 4378/2657 \quad = 39.2 + 0.7 = 39.9 \; mm$$

Total deflection of beam $= 39.9 + 5.7/5.0 \times 34.2$

$$= \quad 78.9 \; mm \; (span/190) \qquad\qquad\qquad\qquad \sim not \; acceptable$$

However, the top chord will reduce imposed load deflections by about 5%. This calculation nevertheless shows that the total deflection of stub girders is often the limiting factor. The chords would need to be significantly heavier in order to reduce deflections due to self weight.

	Job No:		Page *24* of *24*	Rev
The Steel Construction Institute	Job Title *Appendix C - Stub Girders*			
	Subject *Design Example*			
Silwood Park, Ascot, Berks SL5 7QN Telephone: (01344) 23345 Fax: (01344) 22944	Client	Made by	*RML*	Date
CALCULATION SHEET		Checked by	*JWR*	Date

NATURAL FREQUENCY

Self weight + dead load + 0.1 × imposed load (excluding partitions)

$$= \quad 2.7 + 0.7 + 0.1 \times 4.0 \quad = 3.8 \; kN/m^2$$

$$= \quad 3.8 \times 12 = 45.6 \; kN/m$$

Deflection due to these `permanent' loads applied to the composite section (allowing 10% increase in stiffness for dynamic effect)

$$\delta_{sw} \quad = \quad \frac{5}{384} \times \frac{45.6 \times 15^4 \times 10^9}{205 \times 5817 \times 10^6} \times 0.9$$

$$= \quad 22.7 \; mm$$

Natural frequency of stub girder $\quad = 18/\sqrt{\delta_{sw}}$

$$= \quad 18/\sqrt{22.7} \quad = 3.8 \; cycles/sec$$

This is greater than the absolute minimum value of 3 cycles/sec. Full analysis of the response of the floor may be carried out in accordance with reference (16), but the vibration response will be acceptable for normal office usage, given the large area (and hence, mass) of the floor that would need to respond to any impulsive action.

CONCLUSION

The design is limited by serviceability criteria and the minimum spacing of the shear connectors on the stubs.

References

1. British Standards Institution (1994) *structural use of steelwork in building*. part 4: *code of practice for design of composite slabs with profiled steel sheeting*. bs 5950, bsi, london.
2. British Standards Institution (1990) *Structural use of steelwork in building*. Part 8: *Code of practice for fire resistant design*. BS 5950, BSI, London
3. British Standards Institution (1990) *Structural use of steelwork in building*. Part 3: *Design in composite construction*, Section 3.1: *Code of practice for design of simple and continuous composite beams*. BS 5950, BSI, London.
4. British Standards Institution (1990) *Structural use of steelwork in building*. Part 1: *Code of practice for design in simple and continuous construction: hot rolled sections*. BS 5950, BSI, London.
5. Mullett, D.L. (1992) *Slim floor design and construction*. The Steel Construction Institute, Ascot, Berks.
6. Mullett, D.L. & Lawson, R.M. (1993) *Slim floor construction using deep decking*. The Steel Construction Institute, Ascot, Berks.
7. Lawson, R.M., Mullett, D.L. & Rackham, J.W. (1997) *Design of asymmetric Slimflor® beams using deep composite decking*. The Steel Construction Institute, Ascot, Berks.
8. Mullett, D.L. (1997) *Design of RHS Slimflor® edge beams*. The Steel Construction Institute, Ascot, Berks.
9. Nethercot, D.A., Salter, P.R. & Malik, A.S. (1989) *Design of members subject to combined bending and torsion*. The Steel Construction Institute, Ascot, Berks.
10. Baddoo, N.R., Morrow, A.W. & Taylor, J.C. (1993) *C-EC3: Concise Eurocode 3 for design of steel buildings in the United Kingdom*. The Steel Construction Institute, Ascot, Berks.
11. Association of Structural Fire Protection Manufacturers and Contractors (1992) *Fire protection of structural steel in buildings*, 2nd edn revised. ASFPCM/SCI/FTSG (now ASFP).
12. Ogden, R.G. (1992) *Curtain wall connections to steel frames*. The Steel Construction Institute, Ascot, Berks.
13. British Standards Institution (1994) *Eurocode 4: design of composite steel and concrete structures*. Part 1.1: *General rules and rules for buildings*. DD ENV 1994-1-1, BSI, London.

14. British Standards Institution (1985) *Structural use of concrete*. Part 1: *Code of practice for design and construction*. BS 8110, BSI, London.

15. Lawson, R.M. (1989) *Design of composite slabs and beams with steel decking*. The Steel Construction Institute, Ascot, Berks.

16. Wyatt, T.A. (1989) *Design guide on the vibration of floors*. The Steel Construction Institute in association with CIRIA, Ascot, Berks.

17. Ward, J.K. (1990) *Design of composite and non-composite cellular beams*. The Steel Construction Institute, Ascot, Berks.

18. SCI (1992) *Design of composite trusses*. The Steel Construction Institute, Ascot, Berks.

19. Owens, G.W. (1989) *Design of fabricated composite beams in building*. The Steel Construction Institute, Ascot, Berks.

20. Lawson, R.M. and McConnel, R.E. (1993) *Design of stub girders*. The Steel Construction Institute, Ascot, Berks.

21. Brett, P. and Rushton, J. (1990) *Parallel beam approach – a design guide*. The Steel Construction Institute, Ascot, Berks.

Index

accommodation of services, 132, 133
anchorage reinforcement, 101
Asymmetric Slimflor Beam, 60, 215–24
  benefits, 60
  bond stress, 72
  construction details, 63
  construction stage loading, 65, 66
  elastic properties, 76
  fire limit state, 68
  section properties, 64
  section sizes, 61, 62
  serviceability limit state, 67
  transverse reinforcement, 74, 75

beam propping, 165
biaxial stress effects in the flange plate, 42,
  43, 93, 94
bond stress, 72
bottom chord design, 141
Bow's notation, 124
buckling
  lateral torsional, 38, 93
  parameter (u), 39

cambering (cellular beams), 169
cellular beams, 165
  benefits of, 168
  cambering, 169
  geometry, 166
cladding details, 85–9
composite beams, 11, 48
  analysis of, 14, 185–200
  construction condition, 12
  construction loading (0.5 kN/m^2), 12
  effective breadth of slab, 12
  fabricated, 129, 130
  moment capacity, 49
  plastic analysis, 13, 14
  profiled decking, 11
  propped and unpropped, 13

composite slabs
  design, 9, 108
  diaphragm action, 11
  fire condition, 11
  using deep decking (SD225), 103
composite trusses, 112
  analysis, 120
  Bow's notation, 124
  deflection, 127
  layouts, 115
  member types, 118
  moment capacity, 121, 122
  systems, 113
  truss configurations, 116, 117
concrete densities (wet/dry), 7
concrete type and grade, 6
connections, 3, 55, 128, 129, 130, 160
  testing, 10
construction
  condition (stub girders), 150
  condition (trusses), 126
  loading, 7, 65, 66
  RSHFB, 81, 81, 82, 84
  stage, 37, 104, 105, 106
  SFB, 31, 32, 33, 34
  torsion, 44, 45, 46
cube strength, 7

deck shape
  composite beams, 24
  influence on shear connectors, 18, 19,
  20
deck testing, vacuum rig, 8
deck types, 5
  dovetailed profiles, 5
  trapezoidal profiles, 6
deep decking
  instruction notes, 110, 111
  section properties, 107, 108

deflections, 99, 127, 150, 151
  differential, 1, 60
  limits (decking), 8
degree of shear connection, 51, 98, 140
design assumptions
  Slimflor construction, 37, 41, 90
diaphragm action, 3, 11
dynamic sensitivity, 25

edge beams, 54
effective breadth of slab, 12
elastic section properties
  Asymmetric Slimflor Beams, 76
  composite beams, 21, 22, 23
embossments, 9
end diaphragms, 29
equivalent slenderness (λ_{LT}), 38

fabricated composite beams, 129, 130
  accommodation of services, 132, 133
  beam shapes, 132
  floor layouts, 131
fire protection, 103
fire engineering
  Asymmetric Slimflor Beams, 68
  composite beams, 25
  composite slabs, 11, 25
  RHS Slimflor edge beams, 102
  Slimflor beams, 29
flange plate
  biaxial stress effects in, 42, 43, 93, 94
flange ratio (N), 39
formulae, derivation of
  Asymmetric Slimflor Beam, 215
  Composite beam with shallow profiled
    steel decking, 171
  RHS Slimflor edge beam, 225
  Slimflor beam, 201
framing arrangements (trusses), 114
full and partial shear connection, 16, 17, 18

galvanising
  profiled steel decking, 5
gyration, radius of, 40

holes in beam webs, 34

influence of deck shape on shear
  connectors, 18–20
influence of shear (trusses), 143

influence of tension (trusses), 143
installation notes for deep decking, 110, 111
interaction of shear and moment, 15

joint eccentricities, 119
J torsion constant, 40

lateral torsional buckling, 38, 93
longitudinal shear transfer (composite
  trusses), 126

mechanical interlock, 9
mesh, 3
minimum slab depths, 108, 109
minor axis slenderness (λ), 40
modes of failure (composite slabs), 9
modular ratios, 23, 24, 49
moments of resistance
  Asymmetric Slimflor Beams, 69, 70, 71,
    72, 73
  composite beams, 13, 14
  composite trusses, 121, 122
  RHS Slimflor beams, 95, 96
  Slimflor beams, 69, 49, 50, 37, 48
  stub girders, 137, 138, 139, 140

non-composite beams
  Slimflor beam, 41, 48, 233–56
  RHS Slimflor edge beam, 78

parallel beam approach, 156, 157
  beam layouts, 163
  connections, 160
  erection, 162
  fabrication, 162
  framing, 159
  lateral stability, 164
partial shear connection, 51
plastic analysis, composite beam, 13, 14
plastic modulus (S_x), 38, 39
ponding, 7

ratios (span/depth)
  composite slabs, 11
  trusses, 120
RHS Slimflor edge beam, 77
  benefits of, 78
  composite beams, 83, 97, 98, 225–32
  design of, 89

RHS Slimflor edge beam (*continued*)
  non-composite beams, 91
  section classification, 91, 92
  torsional effects, 90
  transverse reinforcement, 100, 101
  vertical shear capacity, 96

section classification, 41, 121
section properties, deep decking, 107, 108
serviceability criteria
  Asymmetric Slimflor Beam, 67
  composite beams, 23
  trusses, 123, 125
shallow profiled steel decking, 3, 171–84
shear connection
  benefits of, 3
  degree of, 51, 98, 140
  forms of, 16
  full and partial, 16–18
shear connectors
  influence of deck shape in, 18–20
  spacing, 52
Slimdek construction
  benefits, 28
  Slimflors, 26, 27
Slimflor beam, 26–111, 201–14
  benefits of, 28, 29
  combination with concrete slab, 29
  design, 36
  project, 34, 35, 36
  test programme, 55–9
  types of SFB, 29, 30
speed of construction, 3

steel grades
  beam sections, 30, 62
  decking, 5
strength design of decking, 7
stub girders, 134, 281–305
  bending moments, 137
  bottom chord design, 141
  concrete flange design, 147
  design considerations, 136
  design procedures, 153–5
  forms of stub girders, 135
  secondary beams, 152
  transverse reinforcement, 148, 149
  Vierendeel action, 137

T section classification, 121
tie members, 31
torsional effects RHS Slimflor edge
  beams, 90
torsion constant (J), 40
torsional index (x), 39
transverse reinforcement, 20, 21, 53, 74, 75,
  100, 101, 148, 149
trapezoidal deck profiles, 6
truss member types, 118
trusses, influence of shear/tension, 143

vertical shear capacity, 96
Vierendeel action
  stub girders, 137, 142
  trusses, 120, 124
Von Mises, biaxial stress formulae, 43

web members for trusses, 119

Learning Resources
Centre